Heidegger and
Ontological Difference

A Rider College Publication

Rider College

Trenton, New Jersey

Heidegger and
Ontological Difference

L. M. Vail

The Pennsylvania State University Press
University Park and London

Library of Congress Cataloging in Publication Data

Vail, Loy M
   Heidegger and ontological difference.

   "A Rider College publication."
   Bibliography: p. 205
   1. Heidegger, Martin, 1889–   —Ontology.
2. Difference (Philosophy) I. Title.
B3279.H49V26   193   70–165361
ISBN 0–271–01108–4

Copyright © 1972 by The Pennsylvania State University
All rights reserved
Library of Congress Catalog Card Number 70–165361
International Standard Book Number 0–271–001108–4
Designed by Glenn Ruby

To George A. Schrader
and Guy W. Stroh

Contents

	Foreword	1
I	Being and the Things-That-Are	3
II	Revealing and Concealing	25
III	Man and the Ontological Difference	47
IV	Destiny and the Ontological Difference	79
V	The Ontological Difference as the Realm of all Realms	123
VI	Language and the Ontological Difference	156
VII	The Continuing Mystery of Ontological Difference	188
	Bibliographic Key to Textual References	205
	Secondary Bibliography	207
	Notes	209
	Index	223

Foreword

Heidegger is well known for his preoccupation with the problem of Being and for the unconventional ways in which he attempts to deal with this problem. These methods are dictated by his conviction that there is a fundamental difference between *Being* (*Sein*) and *that-which-is* (*das Seiende*). Heidegger first used the term *ontological difference* to name this difference in an essay entitled *Vom Wesen des Grundes*. This essay, written in 1928, is one of the so-called transitional essays in that it lies, both historically and in terms of content and spirit of expression, between the stage of his thinking embodied in *Sein und Zeit* (1927) and the later essays.

The difference in style and orientation between the earlier and later phases of Heidegger's thought poses an immediate problem for interpretation. Before it is even possible to deal directly with the problem of the ontological difference, it is necessary to take some stand on the continuity of Heidegger's thought. A detailed pursuit of this question lies beyond the confines of this study; however, Heidegger's later writings are full of suggestions as to what portions of *Sein und Zeit* he considers to have paved the way for problems taken up at a later date. Therefore the fact that the actual term *ontological difference* is not explicitly mentioned in *Sein und Zeit* does not mean that this book is irrelevant to the problem of the present study. Many of Heidegger's own interpretative remarks concerning *Sein und Zeit* indicate that certain problems were implicit in this earlier work and later were made explicit themes for clarification. Of course the problem of proving that something is mentioned *implicitly* still remains. Should we take Heidegger's word for this? This would be merely pitting one portion of his thought against another. Who is to say which portion should have precedence?

I can only propose to answer this question by making clear the intentions of this study, namely to explore the problem of the ontological difference: what the problem is, what significance it has for Heidegger's own thought, and what significance it has as a philosophical problem. Therefore I shall be primarily concerned with the later phases of Heidegger's thought, in which the ontological difference is either an explicit theme of discussion or else is explicitly directive of Heidegger's thought about other matters.

Nevertheless, Heidegger himself is one of the first to say that the way of *Sein und Zeit* is a necessary one for those who would understand his later writings (*Hum* 17; *USp* 137).* I think a good case can be made for the assertion that the two (or three) "phases" of Heidegger's thought are mutually dependent with regard to their interpretation. Although one could do a critical analysis of *Sein und Zeit* without the later writings, I think much of Heidegger's message would be lost, and the converse is also the case. It is in *Sein und Zeit* that Heidegger rigorously sets out what he considers to be the relation between the problems of Being and human existence and the basic methodological procedures dictated by this relationship. Moreover, his later writings are often replies to or discussions of this earlier work, indicating that the work, perhaps not in expression but definitely in meaning—and often in the kind of meanings that have to be read between the lines—has remained for Heidegger himself a focal point for his later thought.

For this reason I will make use of *Sein und Zeit* to the extent that it furnishes background material for the problem of the ontological difference. However, in order to interpret the significance of this material for the central problem of my study, I must rely heavily on Heidegger's own suggestions. The perspective of this study must therefore necessarily shift depending on whether the theme of the ontological difference is pursued implicitly or explicitly.

It would be impossible for me to list the names of all those who have helped me with this work. I will therefore limit my list to four names: Professor George A. Schrader, who was my dissertation adviser and has given me continual encouragement to pursue subsequent work on Heidegger; Karsten Harries, Dominick Iorio, and Theodore S. Voelkel, who read the manuscript and made innumerous helpful suggestions. I would also like to express my appreciation to my colleagues for their diverse ways of aiding this project to its completion.

---

*See Bibliographic Key to Textual References (p. 205) for complete sources.

1

Being and the Things-That-Are

Das Sein ist das Leerste und zugleich der Reichtum ...
(2 *Ni* 250)

Heidegger initially formulated the ontological difference as a difference between Being and the things-that-are. Anything that is, is given as an object which we can encounter or at least exemplify within our world. Some things-which-are are the familiar objects around us; trees, dogs, cooking utensils, automobiles, and so on. Other things-which-are are the so-called sense data; colors, sounds, tactile sensations. Even the "abstract entities" such as the number 3, utility, triangularity, economic man or an n-dimensional figure are things-which-are; although they cannot be directly apprehended, such possibles can be exemplified in or correlated to situations within our world. Finally, even the fictions of poetry and mythology are things-which-are, insofar as they can be conjured up and given definition by the poetic imagination; no one would say that Homer's Apollo is unknowable, and a thinker such as Nietzsche may offer evidence that Apollo is more than a mere fiction. Although Being is "that through which things are," Being is not something-that-is which can be defined or explicated and situated within the world. There are no situations where we can say meaningfully: "Here is Being, there It is-not," as we can in the case of things-which-are. Being is no more and no less relevant in one context than in another. The attempt to discourse directly about Being is either overwhelmed by its vastness—an infinite, inexhaustible totality of things-which-are, stretching backwards, forwards, beyond all possible directions—or drowned in an empty sea in which nothing can be distinguished. Being is, then, transcendent—beyond. Heidegger directs himself to the question:

what sort of "beyond" is involved in the case of Being? What is the meaning of the ontological difference?

Heidegger's formulation of the ontological difference as a difference between Being and the things-that-are is complicated by the fact that Heidegger later gives up the term *Being* (*Sein*) in trying to formulate his own thinking, especially in connection with such areas as thinghood and language. He comes to view himself as going along a path where Being per se is hidden or perhaps has even run through its course. He argues that the traditional views of *Being* have forgotten the difference between Being and what-is, and that this difference only comes to the foreground when these traditional views are overcome. Richardson points out that the course of Heidegger's thought shows a growing suspicion of the term *ontological*, so that his later works refer to *the Difference* rather than to *the ontological difference*.[1] I will argue that to fully understand the Difference it is necessary to go beyond the vocabulary of *Being* (*Sein*) and *things-that-are* (*das Seiende*). But in order to go beyond something, we must first pass through it.

I do not intend here to trace through Heidegger's historical development of these ideas. Rather, what is intended is to see the Difference in terms of the tension between Being and what-is and to see where this tension itself must lead us. References to these terms occur all over Heidegger's thought, and it is evident that Heidegger never completely abandoned his interest in an *ontology*, although his continual reflections on his initial program led him to the view that an *ontology* would only be meaningful if predicated on a more fundamental encounter between man and the primordial, hidden realm which nourishes his thought. He strongly adheres to certain points throughout all his work, and the present chapter will concentrate on those points of commonness. One of these is that the success of any encounter with *Being* depends upon whether Being can manifest itself of its own accord and not merely as the content of our own spontaneous subjective thinking or as the form of empty tautologies. Only if this is the case can Heidegger escape a subjective arbitrariness of a formal vacuousness. Yet what can be taken as valid evidence that Being manifests itself? It cannot manifest itself as an object if it is different from the entities. Pure withdrawal (*Entzug*) can possibly be experienced, but we cannot at the outset distinguish it from an encounter with Nothingness. Furthermore, withdrawal accounts only for the transcendent aspect of Being. If Being has no other aspect than a transcendent one, it is hard to see how the problem of

Being could really be lively and relevant to us. Is it possible for man to have any sort of positive encounter with Being in its difference from what-is? Can the ontological difference itself be encountered, or can it only be spoken about? Or is speaking about something a kind of encounter in itself?

The meaning of Being

Although the term *ontological difference* does not appear as such in *Sein und Zeit*, Heidegger tells us that Being (*Sein*) is not something-which-is (*ein Seiendes*) (SZ 4). Indeed, we can say that this very book was prompted out of an attempt to find a fruitful way of raising the question of Being so as to avoid the impasses of previous ontologies. It is Heidegger's contention that Being cannot be directly apprehended, for we possess no "intellectual" intuition of a cognitive sort, although we do have what might be called a precognitive understanding of things-which-are in light of their Being. At any rate, Being cannot be encountered as an object, for there is a difference between Being and what-is. Yet Heidegger does not at this point focus on the Difference as such. Rather, he is more concerned to focus positively on the relation between the things-that-are and Being, and also to define the position of the human inquirer and show how his position colors the very nature of an inquiry into Being.

The term *ontology* is a problematic one in Heidegger's thought, as has already been noted. Initially Heidegger used the word to mean an inquiry into the Being of the things-that-are (*Sein des Seienden*). The purpose of such an inquiry would be to shed light on the things-that-are. Yet this endeavor is open to a number of interpretations, and two meanings have prevailed especially: (1) an inquiry into the superlative entity, which "has the most" or best exemplifies Being, and is therefore taken to be the proper starting point and also the proper ending point of ontological inquiry; and (2) an ambitious attempt to construct a system of categories which would schematize all-that-is in light of its ultimate ontological significance. Yet, since the purpose of either type of ontology was to shed light on the things-that-are, such ontologies were really attempting the work of a science.

Heidegger, on the other hand, has claimed from the very beginning that he was interested in bringing to light the meaning or sense (*Sinn*) of Being per se. Thus there is a sense in which Heidegger never was merely an ontologist. In *Sein und Zeit* he keeps reminding us that his primary problem is an interpretation of the meaning of Being. What, then, led him to the idea of a fundamental ontology? Why did he not immediately see that his aims were not those of an ontology in the usual sense?

The reason is twofold. First, he claims that, although Being is different from the things-that-are, the former is not severed off from the latter. Being is always the Being *of* the things-that-are (SZ 6). Were this not so, the problem of Being would not have the central importance it does; for if Being were off in a realm by itself, it might be possible to escape its hold and influence. Heidegger argues, however, that the problem of Being is ultimate and unavoidable (SZ 9). Second, he tells us that all ontology must remain blind unless we first bring to light the horizon in whose light the various things-that-are are categorized (SZ 11). This horizon is not readily manifest and needs to be lighted up by an inquirer; for it is man who plays a crucial role in bringing the horizon of Being to a point of disclosure. This must be so, for otherwise there would be no *problem* of Being—the meaning of Being would be evident directly and immediately to all.

Even though Heidegger later grew suspicious of the term *ontology*, his concern for the Being of the things-that-are has in a sense remained. The later problem for him is, indeed, to overcome the separation of Being and the things-that-are which arises through forgetting the meaning of Being, and to deal with the alienation which grows out of such a separation. But this is a problem for later discussion. For now the question must be, how does the ontological difference influence the possibility of inquiry into the meaning of the Being of the things-that-are (*das Sein des Seienden*)?

It may be helpful to note that in *Sein und Zeit* Heidegger drew a distinction between meaning as sense (*Sinn*) and referential meaning (*Bedeutung*) (SZ 51, 324). A hammer, for example, has referential meaning; it belongs to a functional complex. To understand *what* the hammer is we must understand *how* it fits into this complex. In fact, Heidegger maintains that the order of encounter is such that we first discover an object as a hammer through singling it out of the functional complex. But a hammer is not the sort of thing to have "sense"; Heidegger maintained in *Sein und Zeit* that only Dasein (and the *Existenzialien*) "has sense." Sense involves an awareness of a horizon of projected possibilities—a transcendental region which gives things their orientation and makes them intelligible in terms of their possibilities. "Sense is that, wherein the intelligibility of something is situated" (SZ 151).[2]

A question about the sense of things could never be answered by listing their referential meanings; the latter presuppose the former. Without a predisclosed horizon or region of sense, referential meanings would be blind and arbitrary. When philosophers speak of the *absurd*, they are referring to the alleged senselessness of things, not to a lack of referential meanings. No one would deny that there are hammers pounding nails or that $2 + 2 = 4$ in an absurd world. Rather, what is meant is that the transcendental horizon is closed; hence the various referential meanings seem blind and alien. We can see that an acorn "means" a future oak tree or that the oncoming dark clouds "mean" rain, but why are there such things in the first place? The "why" of things could only be appropriately met with on the level of a transcendental horizon—the level of sense. The transcendental horizon is a horizon of Being.

Although Heidegger later abandoned such terminology, this kind of distinction can still be seen in his works. The idea of an opening or illuminative clearing (*Lichtung*) is present throughout his work. In fact, his very latest works speak of a Region of encounter (*Gegend*) which is the source of presence as well as of the withdrawal of presence. The main difference between the earlier and later works is the position of man. In *Sein und Zeit* Heidegger continually cautions us against identifying *Dasein* with man; yet he himself often inadvertently falls victim to such an identification. In speaking of *Dasein* as "projecting" possibilities, we cannot help but think of *Dasein* as some sort of agent. Thus the Heidegger of *Sein und Zeit* never frees himself entirely from the suggestion that the meaning of Being is to a certain extent a construction—that *man* is the basic meaning-giver; *Dasein understands* itself, *articulates* itself. Heidegger did at that time distinguish an "authentic" from an "inauthentic" or "everyday" mode of meaning-giving. But it is first with the Hölderlin essays that he considered the possibility that man is a meaning-taker—that he is subject to a *Seinsgeschick* which, far from being a possibility projected by man, actually projects itself *to* man and first allows man to be what he is (Hö 34–37; Hum 25). This is only glimpsed at in the discussion of temporality (*Zeitlichkeit*) in *Sein und Zeit*.

The kind of transcendental horizon that Heidegger has in mind cannot be philosophically constructed out of a vacuum. It is *a priori*, "there from the very beginning." All our thinking is only possible from within this horizon. In other words, our thinking is subject to "principles" which themselves cannot be merely "thought up." The traditional philosophers construed these "principles" as rules or categories which statically determined the way things are for all time. It is against this doctrine that Heidegger is directing himself when he speaks of "projections." There is no guarantee that the horizon will always be articulated in the same way; in fact there is evidence against this, as even a most cursory acquaintance with the history of ideas will show. Indeed, the horizon may vary according to different peoples and cultures. In the West the horizon has been disclosed as a horizon of Being in sharp contrast to Nothingness; this life is real, and the things that pertain to it must be explicated in terms of their *Being*.

Regarding the everyday meaning of *Being*, Heidegger tells us in one of his later works: "The name names that which we mean when we say 'is' and 'has been' and 'is coming.' Everything which reaches us and towards which we reach, goes through the 'it is,' spoken or unspoken. That things are that way—that fact we cannot escape at any time or at any place. The 'is' remains recognizable to us in all its manifest and hidden variations of form" (*KTS* 5).[3] This everyday, prephilosophical understanding of Being is for Heidegger a *factum* (*SZ* 5); it is, as Werner Marx put it, evidence that Being shows itself of its own accord independently of our projected metaphysical schemata.[4] By approaching Being as it shows itself of its own accord or through its own example, Heidegger seeks to show that the problem of Being is neither a pseudoproblem nor the product of a theoretical construction. It is Heidegger's contention that the problem of Being is in fact the most fundamental and most concrete of all problems (*SZ* 9). In *Sein und Zeit* he sought to establish his case by invoking a version of the phenomenological method. The method of phenomenology, which Heidegger later claims made possible his initial approach to the problem of Being (*USp* 122), expresses the Husserlian maxim "to the things themselves!" (*SZ* 27).

The method, as Husserl had proposed it, involved an attempt to bracket off all presuppositions about the existential status of an object —such as real, ideal, illusory, for example—so that the object could be present to us in its pure what-ness.[5] To borrow a term from Gilson, Husserl's method was an attempt to achieve "existential neutrality,"[6] the purpose of which was to allow one to approach the object as it presents itself most immediately to consciousness, as it presents itself, in its what-ness. This bracketing-off of all presuppositions concerning the existence of an object was known as a phenomenological reduction. The phenomenological reduction was not confined to the object

Being and Things-That-Are 9

however; Husserl felt it equally necessary to reduce the subject down to the phenomenon of pure consciousness.

Nevertheless, Husserl's reductions were always carried out within a subject-object schema; for him an object is always an object for consciousness and consciousness is always a consciousness of an object. This maxim of "to the things themselves *in so far as they are objects for pure consciousness*" is from Heidegger's standpoint a serious compromise of the maxim "to the things themselves!"

For Heidegger the Husserlian maxim means "to let be seen that which shows itself as it shows itself through its own example" (*SZ* 34). A phenomenon cannot show itself through its own example if we superimpose the subject-object framework upon it. Moreover, the notion that a phenomenon shows itself as an individual "what"[7] also involves assumptions characteristic of an elaborate philosophical framework. By bracketing only statements about an object's that-ness (*Dass-sein, Existenz, Tatsache*) and not those about its what-ness (*Was-sein, Essenz, Eidos*), Husserl does not in fact achieve a real reduction but rather decides for one group of philosophical statements concerning "Reality" as over against another group. Or he loses the problem of Being altogether in the advanced reductions. As such, his reductions remain within the framework of the metaphysical tradition.

Heidegger carries out the idea of a reduction on a radical level by trying to bracket off all philosophical notions of Reality and Being in order that the phenomenon be allowed to show itself through its own example, in other words, without its character being predetermined through our philosophical preconceptions (*SZ* 16). He takes the meaning of the term *phenomenon* to mean literally a self-showing: "The meaning of the term 'phenomenon' that we should adhere to is: that-which-shows-itself by its own example,[8] the manifest. The . . . phenomena are then the collection of that which lies open in broad daylight or can be brought to light of day; what the Greeks at times simply identified with *ta onta* (the things-which-are)" (*SZ* 28).[9]

A word of caution is needed about the phrase *an ihm selbst*. What Heidegger wants to emphasize here is not some kind of animated phenomena which loom up by their own power; in fact he does not commit himself at this point as to whether these phenomena show themselves by their own power or whether they are uncovered by something else. With regard to the phrase *an ihm selbst*, Heidegger has more in mind a kind of showing which is complete in meaning without reference to something adventitious. He tells us: "There is even the possibility that what-is shows itself as that which in itself it is not. In this self-showing what-is 'seems' or 'looks like . . .' Such self-showing we call semblance or seeming" (SZ 28–29).[10] Thus *an ihm selbst* refers to a showing which is taken as just what it shows itself to be. Such a showing makes no reference to anything other than its own manifestation; through its own example it shows itself to be as it is.

By emphasizing the *an ihm selbst* Heidegger contrasts his notion of the phenomenon with semblance (*Schein*) and with appearance (*Erscheinung*). In semblance a thing can show itself as that which it is not, as when fools' gold shows itself to be gold. The ancients always allied semblance with non-being; in semblance a thing shows itself as that which it in essence is not. Heidegger points out however that semblances are grounded in showings; thus all semblances have a real basis and are to be treated as instances of phenomena along with so-called "real" showings or manifestations of nondeceptive objects (SZ 29, 36). Hence his famous statement: "How ever much seeming, just that much 'Being' " (SZ 36).[11] Thus even if the problem of Being is an illusion, it is nevertheless founded upon something real. The problem is to find out as best we can the kind of showing which gave rise to the problem of Being. To state this another way, the problem is to find the meaning (*Sinn*) of Being.

Implicit in the above quotation is the fact that semblance can be understood as grounded in showings in two ways, although Heidegger does not point this out explicitly. In one sense a semblance is grounded in a showing by the very fact that the semblance itself has to become manifest. Fools' gold must be manifest before deception will occur. In another sense, semblance is grounded in a showing in so far as the thing after whose example something is showing itself must also be manifest. We could not have counterfeit coin without genuine coin; we could not have fools' gold without gold. This latter relationship between a semblance and that which the showing is a semblance *of* allows for the possibility that a semblance, if it can be clearly recognized *as* a semblance, points to the existence of its archetype. This is why Plato thought he could argue for the existence of archetypes or ideas of which temporal objects were only semblances. It should be noted, however, that the semblance and that of which something is a semblance must be manifest through a similar medium. Thus, although a mathematical formula designates a certain curve, we do not say that the formula is a semblance of the curve. The factory whistle might signify quitting time, yet we do not say that a tone is a semblance of an event. Meaning relationships of this latter sort are based on appearance (*Erscheinung*) rather than on semblance.

Heidegger takes pains to distinguish semblance from appearance. Whereas semblance is a showing-itself as something which it is not, appearance is a failure to show itself (*Sich-nicht-zeigen*). All indications, manifestations, symptoms, symbols have the formal structure and basis of appearance (SZ 29). Messenger phenomena announce something which "stays behind" and does not openly show itself. Dark clouds, rushing winds and falling raindrops indicate the presence of a storm. They show themselves, not by their own example, but rather *as* the indications of a storm. What is the storm "in itself"? The storm cannot be simply identified with its messenger phenomena, for they are the results which the storm causes. The storm itself is a system of conditions that *results* in dark clouds, winds, raindrops; the storm is what is "behind" these and explains their presence, in terms of "necessary and sufficient conditions." The "storm itself" is a set of pressure/temperature conditions at a certain time and place.

What is the relation between "high pressure system displacing low pressure system" and the cool, dry wind that brushes past my face and tossles my hair? What is the relation between kinetic energy and the West Wind of Shelley's Ode? Heidegger characterizes the "showing" involved in an appearance as arbitrary, since the relations between messenger phenomena and what lies behind them are often opaque. Although we can locate the necessary and sufficient conditions for dark clouds, wind, rain, we cannot explain *why* we have storms. In this respect appearances are opaque; they have referential significance (*Bedeutung*) but there is no sense (*Sinn*) to them.

From this we can conclude that the relation of Being to the things-that-are cannot be one of appearance. Being does not "stand behind" the things-that-are in the way kinetic energy stands behind the West Wind. If that were the case, there would be no possibility of inquiring into the meaning (*Sinn*) of the Being of the things-that-are. In contrast, both phenomena and semblances lend themselves to some sort of elucidation. But in the case of a phenomenon, which shows itself, why is there any need for an elucidation? What sort of phenomenon becomes a philosophical problem? "Obviously something which first of all and for the most part does *not* show itself; which, as over against that which does show itself first of all and for the most part, remains concealed, but at the same time belongs in an essential way to that which shows itself first of all and for the most part—so much so belongs, as to constitute the import and ground of the latter" (SZ 35).[12] Heidegger goes on to say that the Being of the things-that-are is just such a phenomenon.

What gives Heidegger the right to treat Being as a concealed phenomenon rather than as an appearance? This question is important, because, if Being were related to the things-which-are in the way that kinetic energy is related to the West Wind, then to ask about the sense of this Being would be to pursue a pseudo-question. But how do we know when we are dealing with a concealed phenomenon rather than an appearance?

To deal with this question, Heidegger in *Sein und Zeit* attempted to show: (1) That all intentional behavior of Dasein, no matter how profound, no matter how trivial, is rooted in an inescapable *Seinsverständnis*, an understanding of Being (SZ 5, 147). (2) There is a polarity in *Seinsverständnis*, a movement or alternation between an "everyday" or privative understanding of Being and a visionary moment (*Augenblick*) or "authentic" understanding of Being (SZ 146, 287–88, 336–39). (3) This polarity is bound up with the very character of human existence, which has a tendency to slip away or hide from itself (*Verdeckungstendenz*) as well as a tendency to show itself (SZ 58, 222, 311). In the beginning pages of *Sein und Zeit* Heidegger had put the matter this way: "The fact that we dwell per se in an understanding of Being and that the sense of Being is nonetheless enshrouded in obscurity, proves the fundamental need of repeating the inquiry into the meaning of 'Being' " (SZ 4).[13] Where there is *Seinsverständnis* there is a possibility of elucidation, of philosophical interpretation. In such a case it makes sense to inquire into the meaning of Being: precisely, in this situation where Being is inescapable on the one hand and hidden on the other. This theme remains with Heidegger throughout his work.

What is the status of this *Seinsverständnis*? How do we know that it has any kind of objective validity? Perhaps it is like man's belief in the supernatural, a kind of fiction which man will some day have to rid himself of? Much of *Sein und Zeit* is devoted to making out a case for the inescapable reality of just such an elusive understanding. In the long run it is impossible to forget Being; the phenomena of death, dread, conscience bring us face to face with Being. As to the location of *Seinsverständnis*, it must be admitted that the position of *Sein und Zeit* is ambivalent. Although *Seinsverständnis* is not to be construed as a faculty, Heidegger does argue that it must be rooted in an *existenziell* or factical understanding which goes on at a particular place and time (SZ 312). There is a real dilemma here. If the understanding of Being is not *existenziell*, it is a mere theoretical possibility, a "construct"; in such a case there really is no phenomenon, no showing. On the other hand, to say that the understanding of Being must be rooted in an *existenzielles Seinkönnen* (an ability to be which is *existenziell*) is to construe *Seinsverständnis* as something subjective. Heidegger later attempts to correct this situation: "*Seinsverständnis* does not at all here mean that man as subject possesses a subjective presentation of Being and that the latter is merely a representation. . . . *Seinsverständnis* implies that man because of his very essence stands in the openness of the projection of Being and lives with this understanding of Being taken in this later meaning" (SvGr 146).[14] To clarify the character of the projection, he writes: "Being dawns for men in the ecstatic projection. However, this projection does not create Being. . . . The 'jecting' in projection is not man but rather Being itself, which sends man his essence by sending him into the *Eksistenz* of being-here" (Hum 25).[15] At this point the justification for this position cannot be given. It is nevertheless desirable to state it in order that there be some anticipation of the direction of Heidegger's thought.

To summarize, the meaning of Being is bound up with *Seinsverständnis*. Although this meaning may be vacillating and vague, it can never be totally missing. The problem is not to construct a meaning for Being but rather to bring into focus, and address ourselves explicitly to, the source of the meaning of Being we already have. This was the task of *Sein und Zeit*: to track down the source of the meaning of Being. To do this, he takes *Seinsverständnis* as the starting point for his inquiry.

Many questions remain. How does one "grasp" the meaning of Being? To what extent is this meaning graspable? Did Heidegger expect "clarification" of this meaning in the sense that analytic philosophers are after "clarity"? What was accomplished when the meaning of Being was pushed back to the horizon of temporality? What is the relation between the search for meaning and the search for clarity? It would surely be naive to assume that they are the same.

Phenomenology and hermeneutic interpretation

Seinsverständnis is no ordinary phenomenon to be investigated in the usual way by the usual categories. *Seinsverständnis* is unique to man and singles him out from the other things-which-are; but, more important, *Seinsverständnis* provides a clue to the way man exists vis-à-vis Being. Indeed, it seems a bit odd to speak of *Seinsverständnis* as a phenomenon, something which shows itself. One would think it had more the character of an apprehension. On the other hand, any apprehension must *occur* (show itself) in order that something *be* apprehended. What makes *Seinsverständnis* unique is the fact that in its case the apprehension and the showing form a unity. In later works Heidegger speaks of an inner unity between phenomenon and apprehension (*WhD* 147–49; *ID* 19–32; *USp* 257–66).

This inner unity is foreshadowed by the discussion of *phenomenon* and *logos* in *Sein und Zeit*. *Logos* means for Heidegger *discourse* (*Rede*) in the sense of *apophansis*, letting be seen (*SZ* 32). Phenomenology therefore means: letting-be-seen-that-which-shows-itself (*SZ* 34). The fact that Heidegger discusses the phenomenon in conjunction with *logos* suggests that he saw already during the early stages of his career that *phenomenon* and *logos* presuppose one another.[16]

The plausibility of this can be argued on the level of common sense. It is hard to have a self-showing when there is absolutely nothing else around. We do not show ourselves by going out into the wilderness; rather, we must bring ourselves into the open—some sort of public domain—where we can be seen. We can of course show ourselves to ourselves, but we all know this means letting-ourselves-be-seen-by-our-Self-as-we-show-ourselves-to-our-Self and not necessarily as we show ourselves. At least, it is obvious that the idea of a totally non-observed showing seems in itself contradictory. The operational meaning of this tautology is that we cannot divorce phenomena from our way of approaching and understanding them.

It can also be argued that an apprehension, a letting-be-seen, cannot be totally independent of a self-showing. We do of course draw a distinction between looking and seeing. "Looking" may or may not imply that something is being seen. Yet the looking itself is an occurrence, a phenomenon. Otherwise we would have no experience of unfulfilled lookings.

The hyphenated unity letting-be-seen-that-which-shows-itself is prior to either the showing or the letting-be-seen. The possibility of phenomenology is founded upon this inner unity of *phenomenon* and *logos*. Phenomenology is not for Heidegger merely a formal discipline investigating a subject matter; letting-be-seen-that-which-shows-itself is something far more fundamental than an arbitrary or free act conducted by some inquirer out of curiosity. Whether we are explicitly aware of it or not, we are by our very nature always engaged in letting-be-seen-that-which-shows-itself.[17] This is one reason why Heidegger does not follow the phenomenological approach as a rigorous discipline; for to do so would be to take this inner unity for granted and succumb once more to an epistemology, where a free-floating consciousness investigates an object, and the ontological meaning of the inner unity of these is left concealed (SZ 21–26).

Sein and Zeit deals primarily with the search for the meaning of Being. The phenomenon in question is Being. But Being is always Being of things-that-are, and it is these latter which are readily apparent and accessible to us. They fill our familiar world. But what status ontologically do they have? If by "phenomenon" in the privileged sense of the term, is meant "Being of things-which-are," what about hammers, mountains, roses? If these are not phenomena, what are they? The position of Sein und Zeit is that these things are not self-showings; they are only evident as what they are within a predisclosed World, which is a phenomenon—it is only *within* a World that hammers, mountains, roses can be discovered and encountered (SZ 83–89). Within a work complex, itself disclosed through a world, we encounter something *as* a hammer; within a geographical region, also first disclosed through a world, we encounter something *as* a mountain. But a phenomenon in Heidegger's sense does not show itself *as* something, it simply shows itself; it "is" the showing. We are not comfortable saying that it *is* the mountain which shows itself as mountain or that it *is* the rose which shows itself as rose. There is something opaque or even uncanny about these manifestations; they have causal explanations, but they lack sense (*Sinn*). Attempts to explain the mountain or the rose involve the use of adventitious concepts, as though something else is behind them—not a rose or a mountain at all. It was this sort of suspicion that led Plato and Parmenides to emphasize the disparity between a world in which there are true identities or self-showings and the world of vacillating things around us.

Heidegger later drops the term *phenomenon* in favor of the verbal form *phainesthai* when discussing Being: ". . . Being discloses itself to the ancient Greeks as *physis* . . . The etymological roots *phy-* and *pha-* designate the same thing. *Phyein*, the rising-up or upsurge which resides within itself, is *phainesthai*, lighting-up, self-showing, coming-out, appearing forth" (*EiM* 77).[18] Also to be noted here is a new use of the term *Erscheinen*. Here it means to come out or shine forth, to come upon the scene and leave the scene: "Implicit in the inner character of *Erscheinung* is coming- on and off the scene, the yon- and hither- in the authentic demonstrative, indicative sense. Being is thus dispersed into the multifarious things-that-are" (*EiM* 78).[19] One must distinguish pure showing or emergence as such from *what* (it) emerges as it comes to light *as* at any one time, for example, a mountain, a rose, the Roman Empire. Being is an ecstatic showing which opens up a field of encounter *within which* the things-that-are are let appear forth *as* what we take them to be. Mountains, roses, hammers, the Roman Empire, appear forth in light of their Being.

The "ecstatic" character of Being is worth some explanation. This can best be understood through the connection between Being and *physis* or *phyein*. Heidegger contends that originally *physis* meant an *Aufgehen* which rests or resides within itself. The German word *Aufgehen* cannot be easily translated into English. On the one hand it means to rise up, unfold, sprout. But it also means to evaporate, dissolve, diffuse. For example in arithmetic *aufgehen* means to cancel out and leave no remainder. In *Sein und Zeit* Heidegger speaks of an *Aufgehen in das Man*. This double meaning of *aufgehen*—to rise up and to evaporate—becomes increasingly more important for Heidegger after his essay on the essence of truth. Upsurge or rising-up suggests the idea of field or what Heidegger calls *Ec-stase*, being beside or outside itself. There is something of this lurking in the term *extension*; but those philosophers who have taken an interest in *extension* have generally treated it as something static and quantifiable. A rose, a mountain, a dog, the Roman Empire or even a man would be examples of things brought out into view through this rising-up. These are at the same time examples of things which disappear after a time; this disappearance is rooted in the dual character of *Aufgehen* itself. This same dual character accounts for the coming and going, the "hither" and "yon" of *Erscheinung*.

Although Heidegger does modify his thinking with regard to the relationships between phenomenon, appearance (*Erscheinung*), semblance (*Schein*), the modification is not really a radical one. The fundamental or primal ongoing is *showing* and what this entails: (1) a field or horizon, for example, a world within which specific things can be encountered but also hidden and obscured; (2) the things which are encounterable within the field or horizon as this or that; (3) a seeing or apprehension which fulfills the meaning of showing even if only in a shadowy way.

Seeing or apprehending is not to be construed as a passive standing-by. Phenomenology deals most properly with an initially hidden phenomenon that must be allowed to come forth (*SZ* 35). For example our view is always filled or even cluttered by the innumerable things around us. They get in the way of the pure self-showing, so much so, that we are likely to forget entirely about it. This is one reason why the meaning of Being eludes the everyday world. Thus the role of *logos*, letting-be-seen, is crucial. "The *logos* of the phenomenology of human existence has the character of *hermeneuein*, through which the authentic meaning of Being and the basic structures of Being which belong to human existence are *proclaimed* to the *Seinsverständnis* belonging to human existence. Phenomenology of human existence is Hermeneutics in the original meaning of the word, according to which it means the task of interpretation" (SZ 37).[20]

To Heidegger there is no contradiction in the fact that a self-showing must be proclaimed and let come forth: the "self" or self-showing does not refer to some sort of automated happening, but rather to the fact that the sense or identity of a self-showing is not adventitious to it. When something shows itself *as* tragic, we are not certain whether the tragic character is in it or in us; here our interpretation brings something adventitious to the appearing. Not so with self-showing; here the meaning must come from within the sphere of the self-showing itself. This can only happen where there is an inner unity between *phainesthai* and *logos*.

Heidegger claims that the inner unity of *phainesthai* and *logos* was first expressed by Parmenides, when he said that Being (*einai, Sein*, to-be) and Thinking (*noein, Vernehmen, Denken*) belong in a relation of self-sameness (*to auto, das Selbe*) (ID 18; VA 231–51; WhD 147–49). According to Heidegger almost every subsequent philosopher has had his own interpretation of this statement. The traditional view of truth as the agreement between intellect and thing is based on this statement of Parmenides. Kant expressed this inner unity in terms of a transcendental unity of apperception (KTS 20, 22, 27–34, ZSf 23). The German idealists interpreted this unity as the reflexive relation, although by so doing they attempted to reduce Being to a process of thinking, in which pure Being undergoes and receives successive degrees of articulation and determinacy.

The idealist position is objectionable to Heidegger in three respects: (1) By understanding "the self-same" as meaning sheer identity, where one term can be substituted for another, the problematic character of the inner unity between Being and thinking is suppressed (ID 15–19). (2) By understanding thinking as a process whereby Being receives successive degrees of articulation, *phainesthai* is not allowed to culminate in a letting-be-seen but rather in an unfolding plenitude of entities; thus pure Being comes to be identified primarily with something indeterminate which lacks sense (*Sinn*). Any horizon of the sense of Being is thereby suppressed. (3) The idealist position does not clearly incorporate the essence of a finite thinker; rather the process seems to go on by itself and appears to be unaffected by the finitude of the finite thinker (*ID* 37–45).

Kant, in Heidegger's opinion, tended more to give *phainesthai* and *logos* (*Sein und Denken*) their just due, as for example in the passage (Kant, *Kritik der reinen Vernunft*, A 158, B 197) that Heidegger is fond of quoting: "The conditions of the possibility of experience in general are at once the conditions of the possibility of the objects of experience as well" (*WhD* 148).[21] The immediate force of this statement is that it is impossible to speak of objects outside the context of human rapport with them. Thus the transcendental aspect of Being (*Sinn*) not only makes experience possible but, from Heidegger's point of view, actually reveals the things-that-are.

Kant's statement also points toward a twofold directedness in man—towards the objects of experience and also towards the horizon of transcendental conditions of his own experience. While immersed in the open realm of things-that-are and engaging in rapport with them (*Verhältnis zum Seienden*) man also exhibits an open relatedness (*Seinsverständnis, Bezug zum Sein*) to the horizon which makes such open encounter possible (2 *Ni* 203–9). Were it not for the fact that the conditions of experience are also the conditions under which the entities become present, our ideas could never have objective significance. And were it not for the fact that we are openly related to our own transcendental conditions, transcendental reflexion would be impossible, and there could be no such thing as a "Critique of Pure Reason."

The meaning of Being cannot be passively apprehended. It is hidden, dissembled. Just as *phainesthai* has degenerate modes, so too does *logos*—for example, the stubbornness of common sense, the capriciousness of idle gossip. The hermeneutic aspect of phenomenology emphasizes the need to be attentive or open to the meaning of Being, even though it be hidden. Later it will be seen that when *Seinsverständnis* lives up to its own hermeneutic character—an announcing of the meaning of Being and a decision to live with this meaning—it is authentic. When *Seinsverständnis* gets bogged down in the affairs of the day-to-day world, the meaning of Being can become dimmed but never totally lost. Man can fail to live up to his possibilities with respect to Being, but he can never lose Being altogether.

There is no objective, automatic disclosure of the meaning of Being; but neither is man free to construct an arbitrary meaning. Being announces itself (*hermeneuein*) and we must pay careful heed. To achieve this attunement, to live up to the inner unity of *phainesthai* and *logos*—that is the hermeneutic problem.

One Being, many things-that-are

So far, it has been shown that Heidegger associates Being with showing and what this entails: a world or open realm full of things which are encountered by an existing being who "sees" or apprehends. Showing entails showing-as-this-or-that at some time. Associated with showing is a world full of diverse things: "Implicit in the inner character of *Erscheinung* is coming- on and off the scene, the yon- and hither- in the authentic demonstrative, indicative sense of these terms. Being is thus dispersed into the multifarious things-that-are" (*EiM* 78).[22]

Conversely, no matter how different or unrelated things seem, there is an underlying thread to them. Heidegger first argued this point in *Sein und Zeit* when he affirmed the overriding unity of temporality (*Zeitlichkeit*) despite its articulation into the three *Ecstasen* or modes of future, present and past. There he points out that showing cannot be limited to the now-present; such a view would lead us to the position that everything which is, is instantaneous and there are no enduring relations or unities. But even the most extreme nominalists would be unhappy with this, because such a world would be utterly unintelligible and no claims could be made about it. Heidegger argues that those things which are no longer at hand are revealed in exactly that way *as* things no longer at hand, or they are condemned to oblivion. Similarly, to the extent that it is meaningful to speak of things-to-come, they too show themselves *as* things not yet actual. Were it not for the fact that past, present, future are embraced by a common, synoptic showing, all that we associate with responsibility —promise-keeping, planning, acting decisively—would be impossible. In an instantaneous world there would be no plans or intentions; there would be no care or responsibility. This is why Heidegger attempted to show that the care-structure of human existence has a unity to it which is rooted in the unitary character of temporality. We should also note that in *Sein und Zeit* care is the clue to the Being of human existence.

Sein und Zeit takes the position that human existence has a unity to it which apparently overrides historical or cultural variation, at least in so far as the Western world is concerned. Temporality announcing itself in the care-structure is the clue to the unity of this existence. The position with regard to the unity of Being itself (*das Sein selbst*) is however much more ambivalent. We find Heidegger claiming that the Being of the hammer is its hammering (SZ 69). There are innumerable hyphenated expressions—*In-der-Welt-sein*, *Sich-vorweg-schon-sein-in*, for instance—which Heidegger uses to express what he nevertheless considers to be unitary structures. But hyphenated expressions suggest that these unities are synthetic, or posterior rather than prior to the elements into which they may be analyzed. In short, *Sein und Zeit* fails to present us an understanding of Being as a unity that is prior to the plurality of things-that-are. It may be of interest to note, furthermore, that Heidegger becomes much more cautious later about the scope of the unity underlying human existence. Is it man as such or only Western man that exists under the yoke of Being?

When we think of such things as triangles, stones, various people who are alive today, various people who are no longer alive, the Battle of Waterloo, atomic energy, the generation gap, injustice, it can of course be claimed that all of these are. Yet if this is all that is meant by the unity of Being, we cannot help but think that this sort of unity is trivial. To put the problem more dramatically, both justice and injustice, for example, are. But which would seem to be the more important fact: the underlying unity between justice and injustice or the obvious difference between them? Surely they are opposed, as different from each other as any two things could be. Genocide is obviously an example of injustice, and medical help for the aged is obviously an example of justice. They are worlds apart. And yet, if they are so far apart, why do people disagree so much about them? There are even conditions under which some would waver concerning alleged injustice of genocide. Suppose one race asserted itself so drastically as to make it literally impossible for another to live. How much should one take before one is morally vindicated in fighting back? Suppose one race dedicated itself to another's extinction; how would the persecuted race engender its own survival? This sort of wavering relativity is typical of the things-that-are. There are always contexts in which a given thing appears as something different from its usual character.

In *Sein und Zeit* Heidegger stressed the mode of everydayness as a complete articulation of referential structures within which things are encounterable as this or that. Although this everyday context has a remarkable degree of stability, there is a transcendence to it which, however insidiously, forces us to continually reorient ourselves and appropriate the things around us. For example, we have little idea of what the ancient Greeks used to pick and clean their teeth. We know they did not use an Oral-B. In years to come it is possible that the Oral-B will be a curiosity to anthropologists. The identity of the Oral-B as a toothbrush is relative to a given time. It is also of interest to note that the function which a toothbrush fulfills is also subject to change. There could come an age when there is no need to clean one's teeth by any manual means. The wavering character of things-that-are might be exemplified even more dramatically in the case of such things as fashion reversals or moral codes. Here we discover that what is "in" or "right" at some time—bow ties, the "double standard," individualism—is "out" or "wrong" at some other time. There is variety and plurality to what-is; but at the same time, there is some sort of underlying thread, so that a given thing appears some times as this, some times as that, although to be sure within limits. A barbell is unlikely to appear as a crocodile. But slapping a person's face may appear as a duty or as an act of barbarism; a shoe may appear as an instrument of footwear or a means with which to drive a nail into the wall. This is the mystery. Does something show itself *as* a chair because it *is* one? If so, how are we to understand this "is"? What is gained by such an explanation? "Incidentally, it is no accident that the Greek language speaks most clearly and precisely when it names what we call *das Seiende* in the neuter plural. For that-which-is *is* at some given moment and is thus multifarious; Being is on the other hand uniquely singular, the absolute Singular in its unconditional singularity" (SvGr 143).[23]

Heidegger is calling attention to two points here. Whereas the things-which-are are at some given moment, Being is unique. Things which are at some given moment can be generally repeated at some other moment as well. Even the Battle of Waterloo is not really unique. It was like most battles, although to be sure battles do admit of variety. This is Heidegger's second point. What happens or is at some given time may be "succeeded" by something else at a future time. Looked at over a long time-span, there is a diversity but also a monotony to what-is and what-has-been and what-will-be. In contrast, the uniqueness of Being entails an unconditional unity or singularity; Being cannot be broken up so as to exist now but not then, or then but not now, or even *both* then and now as distinct moments of time. Being cannot "be" at some given time. Hence philosophers such as Parmenides, Plato and the entire Platonic tradition treated Being as timeless. "Being is the ultimate universal which is met with in each thing-which-is, and is therefore the most common, having lost every distinguishing feature or perhaps having never possessed any. At the same time, Being is ultimately unique; this uniqueness can never be approached by any thing-which-is. For everything-which-is that might stand out, there is always something like it, i.e., always something else which is, regardless of how different it may appear. Being however has nothing else like it" (2 *Ni* 251).[24] Thus Being has a dual character. It is somehow "common" to all the things-that-are; they all *are*. On the other hand there is an aloof singularity to Being.

Heidegger deals with the "expansive" or universal character of Being in a section in the second Nietzsche volume entitled "Das Sein als die Leere und der Reichtum" ("Being as Emptiness and Opulence"). Here he points out the commonplaceness of our usage of the simple word *is*; yet behind the seeming uniformity and monotony of this word lies a rich variety. He cites examples such as "The man is from Swabia," "The book is yours," "The lecture is in H-5," "The dog is in the garden." Thus Heidegger is keenly aware of the variety or equivocality of the word *is*—aware that Being "is dispersed into the multifarious things-which-are" (*EiM* 78). Although things may exhibit striking differences, they are all accessible as things-which-are. The "commonness" here is not so much an objective characteristic which threatens to level out differences so that we become indifferent to justice or injustice, for example. Rather, the "commonness" refers to a pervasive accessibility: although justice and injustice are opposed, one is no less real than the other. Both affect us; were this not the case—for example, were injustice less real, then we could afford to be indifferent to it in the long run. The fact that it is very real is what makes its minimization a perennial problem for man.

The aloof character of Being is brought to light when we ask: does any one thing-that-is serve as a totally adequate exemplar of Being? The answer has to be "no"—not even man. For to be one thing at some time and place is not to be something else. Being is however "common" to all—to what-has-been, is, will-be. Yet this radical commonness is precisely the feature which singles out Being as aloof and unique. Although Being is "common" to all things-which-are, it "holds back" its identity in any particular situation. Being itself does not become manifest *as* Being but as an emptiness—a Nothing. Thus there is a transcendent side to Being. Although Being "is" pure self-showing, what gets shown are the things-that-are rather than Being as Being. The showing as such harbors a mystery, a refusal to divulge its innermost character.

The dual character of Being thus parallels the dual meaning of *Aufgehen*, the term Heidegger uses to translate *physis*, which he takes to be the original meaning of Being as experienced by the early Greeks. On the one hand Being opens up a field in which we can encounter something *as* this or that item. But the inner secret of this opening-up is held back, and hence the presence of the various things —stones, trees, man, earth, sky,—always has something ominous to it. It is this which makes the problem of Being more than an academic problem.

At this point it might seem as though there were two "differences": there is a difference between the things-that-are and their Being, and there is also a difference within the nature of Being itself. Are there then two Beings (*Sein*) after all, one that is aloof and transcendent, and one that is ecstatic and all-embracing? Or if two-fold Being is still One, how are we to think of this unity in difference? Finally, how is the duality of Being/things-that-are related to the duality within Being itself?

It is doubtful that Heidegger answers all these questions satisfactorily. They do however lead us into the innermost heart of his thought. To deal with such problems Heidegger in his later works comes to the conclusion that we need to reorient our thinking and the guideposts we would use. In this sense Heidegger's work is revisionary and has a curious parallel to certain strains of Anglo-American linguistic philosophical thought which demand that we develop new ways of talking. But Heidegger, of course, moves in a totally different direction. His first move is to rethink the nature of truth. If our conception of truth is faulty, then our whole encounter with genuine philosophical problems will be misleading and probably opaque.

The need for this can be easily seen if we consider one of the standard objections to talk about any kind of transcendence. If something is transcendent, it escapes our knowledge, and therefore how do we know it is transcendent? The argument concludes that talk about anything transcendent is meaningless, because if we knew something was transcendent it wouldn't really be so. Such an argument presupposes a theory of truth—that to be a significant item of discourse is to be present as an object. Truth is classically considered to be an "agreement" between thought and thing. Being is not a thing; therefore does it follow that talk about Being is meaningless? Heidegger's position here is that there are two possibilities. Either it is meaningless to talk about Being or the above conception of truth is inappropriate. If we conclude that talk about Being is meaningless, then we make assumptions about the source of meaninglessness—that it can be hypothetically stipulated or constructed, for example. Heidegger believes however that our cues for meaning are grounded in the total character of human existence, which is much broader and deeper than logical thought. Being does make claims on man in very real ways; this is a central theme in *Sein und Zeit*. To Heidegger Being is meaningful. The problem is to find an adequate horizon or region in which to experience this meaning.

II

Revealing and Concealing

Dann waltet in der Entbergung ihr Sichverbergen?
Ein kühner Gedanke.
(VA 255)

How do we know that such a horizon is possible? Suppose there were no philosophical tradition. Would we in the twentieth century ever come upon Being as an explicit problem? Would Seinsverständnis be articulate and explicit, or would it be more in a kind of semiconscious, dreaming state? Undoubtedly such questions are impossible to answer, but they force us to think about the context and origin of Seinsverständnis. Whatever understanding we have today of Being is conditioned by a tradition; whatever experience we have today, together with its interpretation, cannot be divorced from the Western philosophical tradition and its beginnings. It is for this reason that Heidegger draws much of his thought from the pre-Socratic thinkers. He believes that they had experiences which were crucial for setting the tone of subsequent ones. An example of such an experience is the double aspect of Being, as will be seen subsequently.

Furthermore, the earliest thinkers connected Being with truth. Both had a double aspect. Heidegger suggests that the ancient thinkers saw error and falsity rooted in the very character of truth itself. Error was for them unavoidable, simply part of being human and therefore also part of Being. This point is brought out in a discussion of Being and Illusion (Sein und Schein): "Since Being, *physis*, consists in offering surface-appearances and views, it is by its very nature necessarily and constantly prey to the possibility of a surface-appearance which covers over and conceals what a thing is in truth, i.e., in unconcealment" (EiM 79).[1] Being, pure showing as such, offers this or that as something to take note of, for example, something as a mountain, as a tree, as a hammer. But what *is* a mountain or a tree or a hammer? The fact that we cannot really answer shows that we do not usually note the mountain or the tree or the hammer in their Unconcealment; rather we tend to catch a glimpse of their surface-appearances and then attach meanings which suit our everyday orientation. Thus the truth of the mountain or tree or hammer remains obscure. Being offers us a view; but our viewing of this view as this or that thing-which-is is always to some extent superficial, since we are not let in on the inner character of the offering itself. This brings us to a central theme in Heidegger's thought, that every revealing is at once a concealing.

Although the ideas of revealing and concealing are fundamental to Heidegger's thought, they are also among the most difficult to clarify. They permeate a variety of levels, so that it is difficult to acquire a feeling for when these ideas are appropriate to invoke and when not. On the one hand they seem to be principles of historical explanation; for example, Being has remained concealed throughout the duration of the metaphysical tradition. On the other hand Heidegger uses these ideas as quasi-logical principles. He rejects predicate logic as a normative paradigm for thought because predicative logic is a logic of revealment alone. It assumes the formal givenness of relations and of possible individuals or constants to which these relations can apply. Predicative logic assumes a static conception of formal being, whereas for Heidegger authentic thinking must address itself to the dynamics of revealing/concealing (*EiM* 91–94). Heidegger thus intimates that revealing/concealing might even dictate its "own" kind of "logic."

The status of these principles, revealing and concealing, is thus puzzling. How inevitable are they? What role does man play? Can man maximize or minimize revealing or concealing? Can there be progressive amounts of these? If x can be said to be "more revealed" than y, does that mean that x is "less concealed" than y? What is it that is revealed or concealed? How does revealment or concealment affect the ontological status of things? Revealed or concealed to whom? How are revealing and concealing related to each other?

Revealing/concealing as truth

Heidegger rejects the propositional view of truth as an explanation of the basic nature of truth. The proposition, "The rose is in the meadow," is "true" only to the extent that it leads us to uncover the circumstance of a rose being in the meadow. The rose and the meadow must be *there*, in the open (*SZ* 219–25). In *Sein und Zeit* and in *Vom Wesen der Wahrheit* (1930) Heidegger argued that any normative conception of truth or falsity is thus grounded in the possibility of things being "in the open." He also argued that man must openly participate in this openness in order to apprehend things as being in the open. Human existence and this openness are in a sense congruous, as we shall come to see.

Why does Heidegger attach so much importance to the idea of "openness"? Why doesn't he simply say that the proposition must refer to the fact of a rose being in the meadow? The distinction between referential meaning and sense is involved once again. What is a fact? To Heidegger a fact must already appear within a horizon of sense in order to be accessible *as* a fact: a fact is something which is in the open so that we must take account of it if we are anywhere around it. A fact is an item which is revealed, having come into the open so that it can be noticed.

How do we establish whether something is revealed or concealed? Does the fact that one person may be aware of the rose whereas another is not mean that the rose is in one case revealed and in the other case concealed? At the other extreme, does the rose have to be "publicly observable" in order for us to say that it is revealed, in the open? Answering "yes" to the first question would be to destroy the very possibility of any objective knowledge. It should be clear that this is not Heidegger's intent; although he is critical of the way in which science is pursued, and although he believes that the scientific perspective falls short of an original and fundamental encounter with Being, he does not embrace a radical subjectivism. On the other hand Heidegger is equally unsatisfied with the criterion of public observability. The reason for this can only be understood when we examine the way Heidegger thinks things come to be revealed.

Suppose we see a man smile; we take what we see as a smile. Suppose a child sees a similar expression on a dog's face. "Do dogs smile?" he asks. Why don't we take the dog to be smiling? We speak for example of "smiling inwardly," where someone is smiling even though there are no visible indications of it on his face. To analyze such a situation, Heidegger would reject the view that we first collect phenomena and then infer the presence of a smile. It might be argued that this is a rationale or model to explain the logical pattern of what happens, but this logical pattern presupposes a mistaken phenomenological model. It assumes that there are unproblematically given sets of data which are first there for us and then we "fit" them into a rational model. To

Heidegger there is no such pure, uninterpreted "given" (SZ 150). In this sense he is like the Voluntarists; nothing has an identity apart from what it is taken *as* when it appears or is revealed. Some sort of apprehension is necessary to fulfill the meaning of a showing. Heidegger differs from the Voluntarists however in that for him showing or revealing is not an act of Will. Showing is neither subjective nor objective; rather both subjects and objects are examples of things that get shown.

If we are able to single out and distinguish items which we encounter, it is because we are already working within an interpretational framework, no matter how vague and shadowy it might be. "That which is circumspectively unscrambled or taken apart in terms of its in-order-to-, as such, that which is expressly understood, has the structure of the *somewhat as somewhat*" (SZ 149).[2] In *Sein und Zeit* Heidegger distinguished two levels of "as-structures," the hermeneutic and the apophantic (SZ 158). *Apophansis* means "letting be seen" the things-which-are. But to be something is already to be in the open (SZ 156, 218). Thus the apophantic level deals with what is already there and attempts to describe, delimit, define, clarify it. This is the level of science, for the most part, and also the level of ordinary discourse.

Suppose we attempt to describe a rose—maybe in order to write a field guide to wild flowers. Or suppose we attempt to define the precise nature of a smile. How do we know where to begin? How do we know enough about roses and smiles in the first place in order to describe them? There is an old joke that Adam and Eve were walking around in the Garden naming this and that. Adam points and says, "What shall we call *that?*" Eve replies, "A hippopotamus." Adam asks, "Why do you want to call it a hippopotamus?" Eve answers, "Because it looks more like a hippopotamus than anything we've seen so far." As Aristotle says, all knowledge depends upon previous knowledge of some sort. This is the level of the hermeneutic "as"—the original level at which things are first revealed and made accessible so that they can be talked about.

The act of naming is to Heidegger most important. It is at this level that original experiences are brought to light. The original names given to things often determine the later apophantic modes of talking about them. Names determine for us a primary set of possibilities, basic categories, within which to deal with the things around us. This is what leads the later Heidegger to lay ever more stress on the role of language.

In the case of the rose, at the hermeneutic level the interpretation of something *as* a rose is concerned with the rose *as* a vestige of *physis*. At the hermeneutic level the original disclosure of something *as* a rose is experienced. Such an experience is rare, and it is usually the poet who is able to have this kind of experience with original disclosures. The everyday existence tends to take the rose as a given, thereby assuming the givenness of a hermeneutic interpretation, and then he proceeds from there. Thus the everyday existence misses the rose as rose in its unconcealment and instead apprehends only its surface-appearance.

To relate this discussion to truth, Heidegger is saying that propositional or apophantic truth presupposes a level of hermeneutic or "pre-predicative" truth. This is the original situation of things being in the open. "Being abides as *physis*. The up-rising/dissolving governance is Appearing. The latter brings forward into prominence. Implicit therein is: Being, appearing, allows to come forth out of concealment" (*EiM* 77).[3] Why does Heidegger insist that the hermeneutic level too is permeated by an "as"? Why doesn't he either embrace a logical construct theory or a kind of Platonism? Both the latter theories make the assumption that there are primordial, self-interpreting "givens" which are in a way absolute starting points for our ontic judgments. Heidegger believes that the hermeneutic level has its own ambiguities however. This is highlighted by the child's problem about the dog. The physical features themselves are not what constitute a smile. Rather, it is the way in which the physical features are disclosed. A smile is literally an opening up, a disclosure—something that only a paranoid would have to "infer" or logically construct. But then there are two possibilities here. Perhaps the child is reading into what he sees on the face of the dog. Or perhaps the adult has lost a kind of closeness and intimacy with the original phenomenon of smiling. In poetry and ancient myth there are references to a smiling moon. Sunshine is often associated with smiling and laughter. These in turn are associated with a radiant face. A radiant child is pleasant and smiling.

To say that these are "metaphors" is to uncritically accept a Cartesian dualism. Such a dualism lies behind the comment that the child has committed a pathetic fallacy. A fallacy is nothing more than a failure to observe certain rules. Thus fallacies can only be established relative to a framework. There are no fallacies in themselves. Something must be construable *as* a fallacy. Thus even the attempt to criticize the way the hermeneutic level is structured is itself conditioned hermeneutically—a frightening thought!

The fundamental character of the "as" structure implies that there are no things existing as distinct, isolated essences.[4] But this means that boundaries between things overlap, that things run together. Things have different shades of meaning and importance relative to different contexts. A rose in a bed of prize-winning hiacinths is a weed. A high-heeled shoe can be a handy tool for driving a nail into the wall —much better than a sledge hammer, although the latter "is" a hammer and the former "is" a shoe, and they are ordinarily revealed respectively as hammer and as shoe.

The fundamental ambiguity of the hermeneutic level of meaning has as an important consequence that clarity is an unrealistic goal for philosophical thinking. It is doubtful Heidegger fully realized this point in *Sein und Zeit*. There must be something about the sense (*Sinn*) of Being which brings about the ambiguity of the hermeneutic level, since Being lets things be as this and that. But then to what extent can the sense of Being itself really be clarified either? What sort of an answer can we rightfully expect to the question, what is the sense (*Sinn*) of Being?

The domain of revealing/concealing

"Where" does revealing/concealing go on and "when"? Is this an event or set of events which takes place at a specific time and location? Are there many revealings/concealings? What is the relation between revealing/concealing and the human situation? In *Der Satz vom Grund* Heidegger discusses Aristotle's distinction between things evident or hidden from their perspective and those evident or hidden to us. Aristotle had in mind the fact that the ontic things around us are directly accessible through the senses; we do not have to strain ourselves to be aware of them. Yet they are opaque, they lack sense; their surface appearances do not explain themselves. On the other hand there are principles that we only notice when we strain ourselves, and these principles not only have sense in themselves, but they make the ontic things around us more intelligible by providing a horizon of meaning for them. This distinction is also implicit in Kant's search for transcendental principles of knowledge; Kant would surely not claim that one must be conscious of these principles as principles when drawing a correct judgment. Yet these principles do illuminate and ground our judgments.

Another way to make this point would be to distinguish between revealing and being revealed. This distinction parallels the earlier distinction between self-showing and what gets shown. It would be possible for an item to be revealing (reveal something) and yet itself not be revealed. Such is the relation which Heidegger claims has persisted between Being and the things-that-are.[5] An ontic analogy might be the situation where ultra-violet light reveals fingerprints on a surface. We see the fingerprints but we cannot see the light. Heidegger takes the symbol of light very seriously—fire, light, lightning reveal in that they open up a domain or space of encounter. Without the campfire, for example, the darkness closes in; the fire "keeps the dark away" and creates a region of brightness around it within which things-that-are can be distinguished, however dimly (Hö 55; VA 275). It would also be possible for something both to reveal and to be revealed; signs fit in this category, as do human existence ("Ein Zeichen sind wir, deutungslos . . .") and language (Hö 35; WhD 6–8; Holz 61). For Kant, space and time might fit into this category, since they are a priori intuitions which in turn reveal spatial and temporal objects. Finally, sometimes items are revealed which themselves reveal little or nothing; they are the "trivia" with which huge portions of our everyday lives are taken up. Yet we should note that although the trivia themselves might reveal little, the *revealing* of the trivia might reveal much. The grasp of this distinction is crucial for understanding how an analysis of everyday human existence can bring to light any ontological matters.

Heidegger uses the terms "revealing" and "concealing" to express the inner unity of *phainesthai* and *logos*. For an item to be revealed it must be shown and it must be taken *as* something. Thus the duality of hermeneutic/apophantic is also involved in revealing/concealing. Furthermore, the apophantic level appears to be a double-edged sword. On the one hand it brings into focus the things which are revealed as this or that. But in doing so it makes them accessible in a treacherous way. I can deal with a hammer as though it were a thing in itself with a hard, fast identity. I may become so intimate with the hammer that I begin to wonder where I leave off and where the hammer begins as a thing distinct from my relation to it. I may begin to ask philosophical questions such as "What is a hammer in itself?" I might in this way come up with a theory of Platonic Forms. What has gone wrong? According to Heidegger, hammerness is already an interpretation of what was revealed. I cannot really ask how a hammer in itself came to be; rather, I must "work backwards." I must not ask about the hammer as hammer but about the hammer as a thing-which-is. The truth (unconcealment) of things can only be approached when we take them as things-which-are (das Seiende als solches) and not as uprooted particulars such as hammers, shoes, roses, smiles in themselves (WW 16). The domain of truth is therefore not the particulars themselves but the open domain and the showing which opens up this domain.

What sort of account can be given of scientific causal connections, for example, the connection between dark clouds and oncoming storms or between heat and boiling water? Heidegger's remarks about such things are somewhat sparse and sketchy. On the one hand nothing can automatically be a sign of another; something must be taken as a sign. Yet the failure to take something as a sign is itself an event, and we realize this when we note that a surprising event need not have surprised us if we had been more alert to what was leading up to it. Such connections then depend upon our ability to participate in the open relatedness of things, where everything is set in a context of its possibilities. To grasp water as water is to be aware of at least some of its possibilities and to be able to discover new ones. If a dog is revealed, then an animal is also necessarily revealed. But both revealments occur within an open field or openness which must somehow "contain" already revealed the possibilities of dog and animal (WW 18–19). Without openness per se, this or that item or the ontic relations between them could never be in the open, could never be revealed (WW 11).

What about this openness? It might remind us of the void or "space" of early Greek thought. The scientific mind is wont to dismiss the idea as a kind of mythical hypostatization. However, this may be because we have never clearly understood the function of a "void." Heidegger goes to extensive pains to argue for the necessity of an openness and also to clarify its role. Pure empty openness, if we could imagine such a thing, would be radically implosive on the one hand, tending to surround all things and draw them into it, and yet elusive on the other hand, tending to withdraw beyond all boundaries. This parallels the dual character of physis which was earlier discussed. While withdrawing itself, openness "makes room" for the things-that-are, so that they appear as well defined. It also makes possible ecstatic forms, bounded fields within which things can occur in light of their possibilities. This last feature is an important one, because it emphasizes the interconnectedness of those things which appear within the openness. Threeness and animality are not timeless, irreducible identities; anything which is, has come to be. This means that it has come out into the open, taken its place among those things which are open and evident. The rose has emerged on the scene; were this not so, I would never be able to encounter it; nor would the rains be able to nurture it and pelt its petals. But roseness, that is, to-show-itself-as-rose, has also emerged on the scene, and it is only in light of this that we can encounter this or that as this or that rose. Once again we meet with the dual level of hermeneutic and apophantic truth.

Revealment (*Entbergung*) is thus linked with appearing (*phainesthai*) and also with letting-come-forth (*logos*). Revealment is, furthermore, some sort of event. We can speak of a before and an after, although we may be unable to locate this chronologically. But more important, there is a sense in which a rose can linger forth out of the past and also a sense in which it is no longer present in the way it was when it flowered forth. If the flower withers and dies, it leaves behind remains or a rose-fruit which holds a promise for roses to come. Thus Heidegger is reluctant to confine the domain of relevant experience to the "now" of present actuality (*Gegenwart*). In fact the latter can be looked at as an apophantic instance only made possible by the threefold ecstatic disclosure of temporality per se (*SZ* 218, 406, 416–22). It is the hermeneutic level of disclosure that is co-extensive with the disclosure of temporality per se—the unity of what has been, what now is and what is coming to be (*Zu-kommen, Zukunft*).

If the things-that-are were merely revealed and set out into the open and let stay there, there would be no problem, philosophical or otherwise. The world would be stable or even static. The past would be as vivid as the present. The future would always hold promises but would never threaten to destroy what is now the case. But the kind of world we live in has a dynamic to it which robs it of its long-range stability. Things which are vibrant and vivid today may be wiped out suddenly by a future event; things out of the past may haunt our age so as to dwarf what would otherwise seem important. Things immediately around us may blind us so that we do not fit them into the larger perspective of what has been and what is coming.

Concealment is for Heidegger not explainable by appealing to "human frailty." It would be impossible for us to make mistakes if there were nothing to make mistakes *about*, nothing to lead us astray. In *Sein und Zeit* Heidegger argued that error, concealment and distortion were part of the total structure of human existence. In that work the interplay between truth and error is sharply exemplified in the discussion of authenticity and fallenness. In *Vom Wesen der*

Wahrheit the emphasis is clearly on the structure of *a-letheia* or truth itself. Here he argues that concealment or untruth is not separate but belongs together with truth (WW 17). Yet despite this apparent shift of emphasis, it is *Sein und Zeit* which gives us the concrete manifestation of error, and it is *Vom Wesen der Wahrheit* which attempts to locate the ontological source and significance of error. Thus *Vom Wesen der Wahrheit* and other subsequent works dealing with truth can be seen to complement the work of *Sein und Zeit*.

Error and concealment must be distinguished. The former refers to a situation or realm in which man finds himself as a result of the latter. This realm or situation is one beset by dissemblances and illusions (*Schein, Anschein*). Dissemblance and concealment are related in that dissemblance is a failure for something to be shown as what it most originally is. The smile we give when we pose for a picture is an example. This failure or lack of an original revealing is only possible if concealment is part of the structure to which revealing belongs.

How do we know when something is concealed? How radical can concealment be? Heidegger wants to claim that both revealment and concealment are equally fundamental and that they belong together. Indeed, if anything, concealment might be the more basic, certainly more basic than any ontic revealment: "Concealment denies to *aletheia* its revealment . . . preserves its most unique feature as its possession. . . . The concealment of what-is-in-totality . . . is older than each and every manifest instance of this and that thing-which-is" (WW 19).[6] But how could we know of such a concealment? In the case of revealment, an item is revealed as something, for example, a rose, a clump of dirt, a rabbit. Yet how could we "know" something which is marked by our very failure to encounter it?

Heidegger's answer is based on his widening of the realm of possible experience beyond the usual notion of objects so as to include a dynamics of withdrawal and absence. There is a dynamic to concealment, and we are drawn into this; we are ex-posed, on the scene, where both revealment and concealment take place. Being aware of concealment is something like being aware of emptiness. The bare emptiness of openness can "draw" on us; anyone who has ever looked into a deep canyon and felt the swirling winds rush by him knows this feeling. Heidegger would ask, why is it possible to have such feelings? Birds don't have them. They must be founded in man's relation to the open realm—a relation which cannot be adequately explained by noting only the makeup of man (the psychological approach); for this would not explain why certain feelings have the objective anchors they do. In contrast to an organism with blind fears, we experience a chasm *as* chasm. There is something about a chasm that draws, and also withdraws. Chasms do not generously reveal and give forth; rather, they swallow things up. Chasms are by nature concealing, and our surface encounters with them tend to be negative and disquieting. The openness drives us away from itself, thus preserving its own Mystery (*Geheimnis*) (WW 22).

We can see now why Heidegger attributes a dual character to openness; openness is revealing/concealing. On the one hand it makes room for, or creates, a space within which entities can appear and be given some measure of free play as well as have definite boundaries. But on the other hand the openness itself can never be filled; it withdraws itself and is thus a kind of open transcendence.

The scene of revealing/concealing is coextensive with the scene of human existence. This means for Heidegger that human existence is permeated by error (*Irre*). But error is not merely a subjective mistake about this and that: "The Vertigo, where man is driven around and away from the Mystery towards the accessible items (commodities), away from a present gadget or concern, on to the next and thus missing the Mystery, is errancy" (WW 22).[7] Error has for Heidegger a kind of epic dignity; a paradigm example of error would be the wanderings of Odysseus. Odysseus does not merely make mistakes; but neither is he a blind instrument of the fates. He participates in error but does not originate it. Where then does error reside?

Error prevails in the at-onceness of revealment and concealment (WW 23).[8] To be in error is to be in a state of forgottenness (*Vergessenheit*). Forgetting involves the at-onceness of revealment and concealment. Forgetting something implies remembering something else instead. A stone neither remembers not forgets; it does not dwell in the at-onceness of revealment and concealment. What is it that man forgets? The Mystery which lurks within the openness; driven away and round and round the periphery of the Mystery, man's attention is taken up with the *things* in the open and he forgets the openness itself. He thus takes things for granted and does not see into the conditions for and the source of their presence. His inquiries lose sight of the original, hermeneutic revealing and become superficial. Opinions spring up to counter or add to other opinions. Sometimes ideas are handed down mechanically by tradition. Sometimes traditions are blindly smashed. Finally man is no longer able to say what is revealed/concealed. Yet error itself—fallenness,[9] vertigo, all that goes with it—resides in the at-onceness of revealing/concealing, the twofold dy-

36   *Heidegger and Ontological Difference*

namic of openness, *physis*. As such, error cannot be corrected in piecemeal fashion by man. Rather it must be recognized as part of our situation of being shut out from the Mystery. As such, error must be lived through.

The "terms" of revealing/concealing

Often when we pursue an inquiry, the terms that we are inquiring into seem to undergo a transformation along the way. We may begin by posing nontechnical questions about everyday objects, tables and chairs, and end by posing technical questions about technical objects, atoms, molecules. The tables and chairs get "left behind" in the very attempt to find a truth about what they are. This is typical of scientific explanation. The table as table, or the chair as chair, appears to be left behind by such a procedure. Thus scientific explanations do not give us a table or chair in its unconcealment; in fact they conceal such objects. One is thus faced with a choice—either commonplace objects such as tables and chairs *have* no real truth (unconcealment); or we must reject the method of science when inquiring into such a truth on the grounds that it submerges the objects in question into a sea of universals, and claim that a chair or a table has a truth which is proper for it. Both positions have their problems. Taken to extremes, the first leads us to put down or deny the richness and variety of the world around us in austere, Parmenidean fashion; the second leads to Platonism and the question that troubled the later Plato, "Are there Forms for hair and mud too?" In other words, does everything have a truth (unconcealment) as what it is?

The everyday world has a plenitude of variety, surprise, discontinuity that makes the mind boggle. The amazing thing is that we are able to make as much sense out of it as we do. Yet despite this, the sense (*Sinn*) underlying the presence of things *as* things-which-are remains hidden. This is especially evident in the case of mundane things such as Oral-Bs or dragonflies. Why are there Oral-Bs? Why are there dragon flies? Either question can be "answered" by citing another thing-which-is. Oral-Bs are because of dirty teeth and because not everybody buys Pyco-Pay. Dragonflies are because of dragonfly eggs. But now a new series of why-questions begins—why dirty teeth, why dragonfly eggs? In the case of the Oral-B we end up in an infinite regress and in the case of the dragonfly, in a mysterious circle. In either case the sense, the original truth or horizon of disclosure is missing. This casts doubt as to whether why-questions of this type can ever lead us to a disclosure at the hermeneutic level of sense. This is why Heidegger turns to art rather than to science to explore further the nature of truth. The crucial thing here is that art allows things to be as what they are in their uniqueness; whereas science explains unique items away into universals and abstractions. The toothbrush gets dissolved into a set of supply/demand equations and the dragonfly gets lost amidst a set of biological data. The question as to whether these really have an original truth must be left open at this point.

In the essay *Der Ursprung des Kunstwerkes*, Heidegger characterizes the relation between revealing and concealing as a contest (*Streit*), and it is art which is able to focus and capture the dynamics of this struggle. Thus, as he puts it, Truth happens (*Geschehen*) in a work of art. Not this or that particular thing which is true, but Truth per se happens (*Holz* 26, 44). In other words, art is one of the avenues of original or hermeneutic unconcealment.

One of the examples Heidegger uses is Van Gogh's painting of a pair of peasant shoes. It is worth noting that Heidegger always chooses poetic, rustic objects for this type of example—a bridge or an earthenware crock, etc. Why does he not choose an Oral-B or an airplane or other objects of modern technology? Why does he not choose a dragonfly?

In the case of the peasant shoes Heidegger claims that Van Gogh's painting discloses these shoes in their unconcealment. What is different about the painting as compared to a pair of shoes literally worn by a peasant? The peasant woman uses the shoes; she consumes or uses them up. When they are worn out, she replaces them. The replaceability of them creates the impression that they are mere things hammered out by an abstract production process and then sold in stores. Thus the shoes are in a way "belittled" or "put down" as commonplace (*Holz* 24). Such is the very antithesis to unconcealment. The painting, on the other hand, allows the shoes to be present in their fullest significance. The various features of the shoes themselves are vestiges of a whole world and way of life. Why are the shoes rustic, why do they show the marks of hard use? Why aren't they lacy with high heels? "In the shoe-thing stirs the reticent call of earth, her quiet bequest of the ripening grain and her unexplained abstinence in the desolate fallowness of the winter field . . . servicibility. But this itself rests in the fullness of an essential Being of the tool . . . reliability. By virtue of it the peasant is, through this tool, let in on the silent call of earth, by virtue of the reliability of the tool she is certain of her world. World and earth are for her, and for those who are in her manner, only there in that way: in the tool" (Holz 23).[10]

The shoes point up the way the peasant woman abides—on the earth, in her world. The earth silently calls, sometimes yielding forth and at other times withholding. Why does Heidegger say that this abstinence during winter is "unexplained"? Surely we all know what is responsible for the change of seasons, and we know that when temperatures fall to a certain point, plant life goes dormant or even dies. Yet, why are earth, sky, life subject to the rhythms they are? Why couldn't there be something that the earth would yield out of the barrenness of a winter field? Why must the winter field lie fallow? The earth harbors secrets which permeate all those things bound to it. Heidegger gives the example of colors and tones; the moment we try to analyze them, we lose them. The earth does not tolerate attempts to open up her secrets (Holz 35–36). Yet neither does earth simply hide. Earth is that which comes forth as that which ensconces in safekeeping (das Hervorkommend-Bergende). Earth comes forth in a diversity of ways and shapes—plants, animals, mountains, rivers, colors, sounds. Earth harbors these while they are, but in the end receives back what it gave forth. Heidegger speaks of *the* earth—there is only one. Earth expresses that common ground or anchoring point of life which cuts across all biological, cultural, historical differences. Yet in another way it is earth which lets such contrasts flower forth. Earth is the simple maintainer and supporter of all that we have and are. Even space programs are earthbound. But earth here is not to be thought of as a planet like other planets. If we were to colonize another planet, that too would be presumably earth. Yet this would be for Heidegger a very ominous example, for colonization of another planet could possibly mean cancerous multiplication of numbers of the human race so that earth is squandered mercilessly. Although earth is not to be thought of in the planetary sense, Heidegger nowhere suggests that it ought to be equated with material resources either. Either characterization of earth looses the uniqueness of earth *as* earth. To Heidegger earth is ". . . that to which the upsurgence/dissolution (Aufgehen) of all that upsurges/dissolves, and indeed, as such, comes back into ensconcement. In the things-that-arise/dissolve earth abides as that which ensconces in safekeeping" (Holz 31).[11] There is thus a connection between earth and *physis*. Earth is the self-concealing pole of *physis*. But earth does not merely conceal itself or we would never know of it; rather, earth shows itself *as* that which is opaque and self-concealing (*Holz* 43). Earth is thus an example par excellence of revealing/concealing. More correctly, earth *is* revealing/concealing.

But earth is only one part of the unconcealment or openness which the painting discloses: "To the openness belong a world and the earth" (*Holz* 43).[12] Man, on the earth, is in a world. The unconcealment of the shoes as tool involves their earthboundness and their worldliness. Both of these tend to be overlooked in everyday experience (*Holz* 24, 59). But to a world that has no room for peasant women—a world where all agriculture has become mechanized perhaps—the disclosure of peasant shoes would lack relevance. We cannot become certain of our world today by wearing peasant shoes; this is the problem with all back-to-earth movements. Earth itself can only be vis-à-vis a world. Concerning world Heidegger says: "World is the perpetual non-objective, that to which we are subordinate, as long as the courses of birth and death, blessing and curse, hold us enrapt within Being. Where the essential decisions of our historical destiny fall, there World governs. The stone is worldless. . . . On the other hand the peasant woman has a world because she abides in the openness of the things-that-are. . . . Insofar as a world opens itself up, all things receive their leisure and haste, their distance and nearness, their breadth and narrowness" (*Holz* 33–34).[13] "World" is for Heidegger neither a logical domain, such as the "world" of physics, nor an individual outlook, such as the "world" of Einstein. Neither is "world" a term used to designate an abstract political entirety. For Heidegger the "world" of World War I and World War II are not necessarily the same "world." Furthermore, World War I is no more "worldly" than the Thirty Years' War or the American Civil War.

On the one hand, world involves a framework of fundamental decisions which shape and articulate our possibilities, values and hopes and organize our life, work and leisure time, the political structure of our society, and so on. These vary from time to time and from people to people. Yet a world is not abstract or "ideal"; values and possibilities have a way of being expressed and articulated in terms of our day to day concerns and cares. If a world has room for gods or heroes, we can expect to find statues, murals, and so on, depicting them and customs and ceremonies that honor them. Are there hospitals? Where are people born? How are they brought up? Are there schools and school buildings? Are there graveyards? What sorts of monuments, if any, which commemorate the dead? Are the monuments uniform? Are they all small? Are they all large? Is there diversity, as though the size of the monument were a measure of the esteem the individual held during his lifetime? Have the gods or heroes declined to be part of that world? Do the monuments reflect a sense of loss or mourning for such gods or heroes—"Wozu Dichter in dürftiger Zeit?" Or perhaps the monuments themselves have lost their sense and have become mere commonplace things for pidgeons to disfigure.

The peasant woman, in contrast to the stone, has a world, because she lives in the openness of the things-that-are. "But," Heidegger tells us, "the world is not simply the openness which . . . corresponds to the illuminative clearing. Rather, world is the illuminative opening-up of the paths or essential lines of direction, according to which all decisions are structured and made fit" (*Holz* 43).[14] World as illuminative clearing is not some sort of universal, timeless set of floating possibilities—an abstract openness. World is more than a possible world. A possible world we are free to reject. However, world in Heidegger's sense *prevails* or governs (*weltet*); its sense appears to be heavily fate-laden or "regulative." In *Sein und Zeit* (135) one of the fundamental features of human existence Heidegger calls facticity or being thrown into a world. There is a ring of destiny here. Why is the world of the Homeric age not open to us today? This is something we cannot really explain. If we could—in fact, if we could really *define* world—it would no longer be something we are subordinate to. We would know its secrets and be able to command them. But man is finite; his control over the terms of his existence will always be incomplete because at any moment man is in a world which is not of his own making. Thus man is at any time always existentially on the defensive; he must address himself to something that is in a way alien to him. We cannot master the given as such; we can only learn to work with it and reap its benefits when the association between man and the given has been a happy one and suffer when this association has been unfortunate. Thus our decisions always require that we keep an eye on "consequences." But again this "given" here is not an abstract, timeless sort of thing; rather, it is a set of existential possibilities, which are relevant to our decisions in one age but not in another. A nuclear war or an inflationary period are things for us of today to worry about; these were not problems for the Greek peasant.

Truth or openness then does not consist of a timeless, "objective" set of facts or possibilities. Truth is a dynamic interplay of revealing/concealing. As such, truth has to happen, take place, be anchored and localized. Truth or revealing/concealing cannot be derived from the everyday world and the things in it; we are in need of a prior horizon which must come from "without." As such, it is a gift, an overflow (*Schenkung, Überfluss*) (*Holz* 59, 62). Truth in its highest and deepest sense is always original; it is founded (*stiften*) through poetic art or thought, not caused or derived from the things-that-are. Truth is not a system of mechanically related parts and wholes but rather a dynamic happening: "Truth is present in an essential way only as the contention between illuminative clearing and concealing in the polar opposition of world and earth" (*Holz* 51).[15] Only at this level do the essential "terms" of revealing/concealing begin to emerge. Earth as essential self-closure can only be earth by protruding into world; world as essential illuminative opening must govern and anchor itself in earth in order to be more than a possible world. World breaks open through art and the artwork. The work sets up a world (*aufstellen*) and it reserves the earth (*zurückstellen*). Both of these require some explanation.

In reserving the earth the work lets the earth be present as such without destroying or consuming it. Earth here includes all that belongs to it that surges up through concealing its inner character. When earth is treated as "matter" it is put down as mere potentiality. This is the very opposite of reserving; for matter is looked upon as having little or no character of its own, whereas earth is the self-concealing upsurge as such, the pedestal of *physis*. In the absence of the artwork and its world, or in the absence of any alternative way of revealing/concealing, earth is lost and all that man acknowledges is matter. The experience of earth *as earth* is rare and precious, as is the experience of revealing/concealing rather than a derivative thereof, such as is pursued in the everyday perspective. For Heidegger great art always has something earthbound about it: the stone temples, the landscapes of Van Gogh, Hölderlin's rivers, etc. A work of great art never subdues the earth but lets it flower forth as the self-concealing pedestal of *physis*. In contrast, a jet plane subdues the earth and makes it disappear; the maintenance of the plane consumes resources which must be procured at the peril of much natural beauty. The current environmental crisis is an ontic working-out of what is a fundamental point of Heidegger's thought.

In setting up a world Heidegger has more in mind than the positing of a framework of possibilities, as has already been seen. "Setting up" in its fullest sense has the connotation of consecration (*weihen*) and glorification (*rühmen*). Whatever is consecrated and glorified is taken out of the usual functional context and set aside as what it is. A shoe or a hammer stabilizes and localizes a world for the peasant woman or the housebuilder, but these do not set up world *as world*, that is, world as pervaded by earth and as orienting itself into earth. Truth as the contention between illuminative clearing and concealing remains at best only in the background.

Although man participates in the happening of the opening up of the polarity of world and earth and in the contention between illuminative clearing and concealing, it would be misleading to say that Heidegger views truth as "man-made." Truth is not arbitrary, even though its source lies "without" the things-that-are. Truth is a gift, a founding, an overflow. But what overflows or what gives, if not man? Isn't it man who makes the work of art? Heidegger stresses the fact that it does not appear to be totally within man's power to bring about unconcealment. Some works of art are "unfortunate." Some human attempts are not art at all. We often say that artists must be "inspired" (*Hö* 43, 66–69).

The source of inspiration is holy and forbidding, illuminating and dangerous—like the sun or lightning. Reckless exposure to them destroys man; but without them his domain would be dark and blind. It has been previously seen that Heidegger associates light and fire with openness, and darkness with closure. Heraclitus likened world and *logos* to fire. Zeus was the lightning god and the most powerful and the rightful lord of the gods. The gods for Heidegger are "they" who dwell in the region of the holy without peril; they are in a sense the voice of the holy, the voice of the source of the essence of things—the illuminative clearing and its concealing. ". . . it is in and through the luster that the god is present. In the luminescence of this luster gleams, i.e. illuminatively opens up, that which we call world" (*Holz* 33).[16] This is why Heidegger associates setting up a world with consecration and glorification; the most original happening of truth would have to turn toward its own source. Because the source is holy, however, it cannot be fastened upon and made accessible as can a thing-which-is. The artist-poet can name the gods but cannot grasp them as instruments to be fit within the functional order of things.

Heidegger's treatment of revealing/concealing is not purely formal in the sense that it applies to just anything. Truth has a crucial content as the contention between illuminative clearing and concealing in the polarity of world and earth. Unless world and earth are let *be*, truth in any original sense cannot take place. This contention must be localized in something-which-is; it cannot occur in thin air. Thus the contention requires the participation of man. But man cannot simply make truth; man must take his orientation from the hidden, holy source. Man is shut out from this source and cannot apprehend it directly. The gods, who dwell nearby the source, give hints to men. But these hints, like all hints, are permeated with ambiguity. It is poetry which most nearly does justice to these hints; poetry is the language of revealing/concealing (Hö 42–45).

Fate and revealing/concealing

Revealing/concealing has been characterized as a happening, a contention. Does the contention ever end? What "happens" after truth happens? Are such questions in order? For Heidegger each age or epoch—for example, ancient Greek, ancient Roman, mediaeval, modern, romantic—means the breaking open of a new world. Each world determines the possibilities for apprehending a thing-that-is as this or that—a river as a tree of life, as a means for steamship transportation, as a source of hydroelectric energy, for example. What sort of transition is there from world to world?

What does it mean for something to "happen"? Here again there is an apophantic and a hermeneutic level. On the apophantic level, happenings are structured according to preset or "programmed" possibilities, for instance, will it rain today? It is on this level that we deal with the innumerable everyday events that come about. They do not involve historical decisions which would drastically alter the essence of things; rather, they are simply the workings out of such historical decisions, for example, will the flight to Frankfurt be on time. "History is seldom. History is only then whenever the essence of truth is decided incipiently" (Hö 73).[17] We tend to think of events as instantaneous, as though a thing happens and then it either changes into something else or continues exactly as it is. We observe for example that a lightning flash occurs at 4:00 A.M. and sets a building ablaze. It takes the firemen until 7:30 A.M. to bring the fire under control. Meanwhile two persons have died. In this way we treat events almost as static objects that can be pinned down with dates. We tend to think that once a thing has happened, it is no longer happening. America, we say, was discovered in 1492. How long did it take to discover America?

On the other hand, any event can be analyzed into a set of subsidiary events, possibly *ad infinitum*. The event of bringing a fire under control has "stages" which in turn have substages. Thus it is difficult to know what is meant by philosophers who maintain that events "succeed" one another. Is there an absolute, discrete unit of happening? Is the coming-to-be of each and every commonplace item around us to be taken as an event? Or if these coming-to-be's are not events, what are they?

This problem has plagued much of philosophy. If the separateness of events is emphasized, such that every event is some sort of original existence, it is difficult to account for the continuity and thread to things. Taken to an extreme, such a view ends up with a chaotic and capricious world. On the other hand, suppose we deny that these innumerable, commonplace things are real happenings. We then run the danger of a monism or at best a system of relations in which there are no real termini. If everything is fated, then to what extent does anything have a character and life of its own?

Does Heidegger have any better luck with these problems than other philosophers have had? By claiming that history is seldom, by implication he is claiming that the commonplace things are not themselves original happenings. Yet we must see that for Heidegger this does not involve questioning the reality of these things. It is more a question of whether they succeed in defining and giving character to the open realm within which we encounter them. In other words, it is a question of their truth.

For Heidegger a happening is a decision (*Entscheidung*). In the contention between illuminative clearing and concealing a rift (*Riss*) is opened up between them (*Holz* 51). This rift however does not separate the two so that the contention ends once and for all; rather, boundaries are drawn so that world and earth can be truly vis-à-vis one another in an intimate way. Through the rift the boundaries are drawn which form the outlines within which particular things can be revealed as this or that. The drawing of such boundaries is then an original happening. But it is also a decision, based on the way illuminative clearing and concealing work out the contention. "The illuminative clearing in which the things-that-are stand into, is in itself at once concealment. . . . the open place in the midst of the things-that-are, the illuminative clearing, is not by any means a rigid stage with the curtain always drawn, upon which the play of the things-that-are plays itself through. Rather, the illuminative clearing happens only as concealing" (*Holz* 42).[18] This is why there is such a disparity between the level of the commonplace and the level of original happening. The contention itself tends to be concealed. When man is shut out from something, he turns away from it. This is how things get neglected, overlooked, forgotten. When illuminative clearing and concealing are de-cided, concealing is itself concealed. This is the primordial happening which characterizes the existence of man: he lives in a region which is opening itself up illuminatively by closing itself off. As such, man's region has a flux or drift to it; there is no stage, no eternal logical space in which a thing may be what it is in a Platonic sense.

Although there has been a decision of the contention, there is a sense in which it is still on-going. An original beginning is never an instantaneous fleeting sort of affair; it is not something which happened long ago at a certain date.

"The true beginning contains already concealed the end" (*Holz* 63).[19] This explains the thread of continuity of the things around us. But this continuity will remain opaque to us unless we attend to the original opening up of the horizon within which these things take place.

In one sense the illuminative clearing is the "same" for us as it was for the Greeks; for our world and their world are predicated on the same original beginning. But this "sameness"—Heidegger later calls it the *Seinsgeschick* (fatedness of Being)—is permeated by a kind of "drift," a transcendence or freedom, due to the fact that openness is never a rigid, static logical space but rather a dynamic on-going in connection with concealment. Man is thus driven out beyond the usual and familiar, uprooted from traditions, allowed to tingle and throb with excitement and adventure or to suffer anxiety and desolation. Man does not comfortably reside at the scene of the contention as though he were watching TV. Revealing/concealing as on-going sends itself (*schicken*) along a course. As such, there are no timeless revealings which give us repeatable truths which we could encounter in their exact sameness over and over again.

To see what Heidegger has in mind here we can use a very simple example. Suppose a world of two objects, neither of which has reproductive powers. Such a world will always have two objects, and they will always be related to one another in the same ways. In such a case it might be possible to have a world with no drift to it, such that it is indeed a rigid stage. Now suppose a world which begins with two objects having reproductive powers. At a given time a third object emerges. This third object completely changes the structure of the open realm. It closes off the possibility of a world that has only binary relations; for any binary relation now must have implications for a third object, and hence the new world must have at least one triadic relation in it. Now suppose that this third object is "taken away." Although the world reverts to a two-object world, it is not the same, original world. The new world has a sense of loss hovering over it. Yet both the possible addition and the possible loss are implicitly "given" in the beginning. In this sense the beginning has already overtaken this or that possible happening.

Because the course of the open realm is self-concealing, no formula can ever grasp it as a totality. There is no universal method or universal pulse of things that man can grasp. Man is compelled to wander along with the drift of the illuminative clearing and its concealing. This is the significance of being shut out from the self-concealing, holy source. The story of man is the story of his wandering in the drifting openness and closure of the open realm. But this is the story of Being: the original de-cision where a world broke open protruded by the earth—where the things-that-are came into prominence and the open realm and its dynamic, where the contention between illuminative clearing and concealing remained concealed. This is the way truth originally happened, and we are still working out the fateful course resulting therefrom. It is for this reason that Heidegger considers metaphysics the history of the concealment of Being (2 Ni, 355, 370).

III

Man and the Ontological Difference

Wenn aber das Sein in seinem Wesen das Wesen des Menschen braucht?
   (*Holz* 343)

The inner unity of *phainesthai* and *logos*, self-showing and letting-be-seen suggests that revealing/concealing does not play itself through before a passive audience of men. Were the ontological difference merely a theoretical distinction which we made in order to be able to more clearly organize what we know and encounter, there would be no need to make the relation between man and the ontological difference a special topic for discussion. However, the very context in which the ontological difference is brought up by Heidegger, both in his earlier and later works, suggests that man is bound up with the ontological difference in a most problematic way. In the 1928 essay that first makes use of the term *ontological difference* he wrote: "If however human existence owes its distinctive character to the fact that, understanding Being, man has rapport with the things-that-are, then *this* ability to differentiate, through which the ontological difference becomes factical, must have anchored the roots of its own possibility in the ground of the essence of human existence. This ground of the ontological difference we call the *transcendence* of human existence in anticipation of what is to follow [in this essay]" (*WG* 15–16).[1] Twelve years later in the second Nietzsche volume in a section with the heading, "The rapport with things-that-are and the affinitive link to Being—The ontological difference,"[2] he writes: "We stand in the differentiation of things-that-are and Being" (2 *Ni* 207).[3] Then, finally, in his dialogue between an inquirer and a Japanese (1953–54) he refers to man as he who is used by the twofold (Heidegger's later name for the Difference): ". . . the twofold itself first unfolds the clarity, i.e. the illuminative clearing, within which it first becomes possible for man to distinguish things-present as such and presence... for man, who by his very essence stands in an affinitive link with, i.e., in the usage of, the twofold" (USp 126).[4] Thus, throughout his entire work Heidegger views man as caught up in the tension of the Difference. Man abides among the things-that-are but man has a special link (Bezug) to Being.

The early passage suggests that it is through man that the ontological difference becomes "factical," that is, the Difference is "in" man. The later passages emphasize that man is "in" the Difference. On the one hand the ontological difference is thus constitutive of the very being of man; on the other hand man is the crucial thing-which-is for whom the ontological difference prevails. It is only in contemplating the way that man exists that the Difference becomes relevant; but if this is so, what evidence do we have that the Difference is in any way "beyond" man so as to be constitutive of man rather than derivative from man?

This question brings us back to the subject of the fundamental ontology and shows that no interpretation which tries to explicate Heidegger's later works can successfully circumvent *Sein und Zeit*. For it is in this work that Heidegger presents the evidence that man's existence is characterized by a twofold tension: preoccupation with the things-that-are, both objects and men, and a concern for the meaning of Being.

The transcendence of everyday human existence: man as being-in-the-world

The ontological difference becomes "factical" through the ability to differentiate, which in turn is rooted in the *transcendence* of human existence (WG 16). Heidegger takes the term "transcendent" very literally, suggesting that it means a going-beyond. As long as human existence *is*, it is fundamentally a going-beyond. This means that at any particular moment, man has already transcended something but that there is also something that he has not yet gone beyond. Otherwise man would not be presently *going* beyond. As Karsten Harries points out, "*Dasein* is indeed not a fact, but a nothingness: a relation, a gap, an in-between."[5] There is something which man is beyond, but there is also a "beyond" which holds sway over man. However, this latter "beyond" is not a thing-which-is; it is rather the open horizon itself with its twofold character of revealing/concealing. Thus, man is *between* the things-that-are, which he is also beyond, and the open horizon of Being, which is beyond man, in that it withdraws and eludes his grasp.

Although man is in the midst of things-that-are, he is also beyond them. Man is thus the scene of a dynamic tension which Heidegger characterizes as factical existence. We are factical in that we are involved in an immediate, concrete situation whose conditions were not laid down by us, but given by a past which is with us. Our facticity is a mark that a claim is laid to our essence such that we are not something free-floating. We are not "free beings" in the Enlightenment sense but are rather attuned and conditioned to correspond openly to the open realm (world) in which we reside. Our very residing in the world is not a voluntary act; we are "already here" ever since we can remember, and we "have to be" here as long as we are (SZ 135–37).

Although we are factical, we are also "out beside" our factical situation; we are ecstatic beings (SZ 191, 364–66). This means we can think of distant things and events; we can thus occupy different modes of distance and time. We can live in the past, present or future. We can be a million miles away from something. In this way we have access to and can traverse the whole of the horizon of disclosure, the open realm or world. It is this ability to be ecstatic—to participate in the openness of the open realm—which allows to be "beyond" the things-that-are and to order and arrange them, to be calculative (SZ 220; SvGr 167). If a dog gets his leash tangled, he is totally a victim of circumstance. He cannot view his situation as a whole but merely uses brute force and instinct, attempting to break free. Often this merely makes the entanglement worse. A man, when confronted with an unexpected problem, can project himself away from it and thus get a bearing on the problem as a whole. He can determine causes, appropriate tools, plan for future prevention, and so on.

We are not part of a work context in the same way as a hammer; the latter acquiesces into a fixed, limited function and can only be brought to bear on that small sphere of influence. The hammer does not dwell; its relation to the context is solely one of categorical relations. A man may be assigned a certain portion of a work context, but his relation to the context is never merely that of a tool. The man can view the context as a whole; he can survey and coordinate the categorical parts of his work context. He can select tools and make evaluations. Man dwells by his work; he experiences pride, satisfaction, frustration, fatigue, boredom, and so on. The problem of Verdinglichung could never arise if man were noncalculative and totally acquiescent—if he were merely a tool and not a tool-user. But man is also much more than a tool-user; it will become evident that it is not our calculative nature which corresponds most adequately to our ecstatic dwelling.

Factical existence expresses for Heidegger a twofold movement, involving an ecstatic going-beyond and then a returning to our factical situation, whereupon we encounter the things within the world (SZ 366). But this dynamic, reflexive relation should not be construed as an oscillation of the metaphysical kind, involving the discrete poles of material actuality and transcendental categories. Our return to the factual situation is never merely a return to material actuality; rather in returning we take up residence in the proximity (Nähe) of that which seems to lay claim to our essence[6] (Gel 68, 72–73; ID 45–47; SvGr 158–60). Factical existence sums up the necessary and finite character of man's essence. As factical we are limited and contingent, but not accidental; we are attuned, and this means that we belong openly and necessarily in the world.

Our everyday mode of existing is primarily oriented towards the immediate things around us rather than towards Being per se. Nevertheless, we can understand, evaluate, use and care for these items in terms of their Being. Items cannot in and of themselves be given for consciousness, nor can there be a plenum whose givenness is nonproblematic. The fact that we are drawn in by the thing-world and that we become so involved with it shows that the givenness of things is not a mere thereness which we can take or leave. One of the very conditions of our existence is that we reside in the midst of things. Things "demand" our attention; they sustain and threaten, give pleasure and pain, distract and attract. The very givenness of these things is highly problematical (SZ 136–40).

Abiding in the world, we abide in the midst of things; but things are only so far as they are within the world. Thus no entity truly is in itself but takes on its very character only in view of world per se, which bounds, defines and illuminates the things and thus gives them over to us (SZ 55–57). To view an entity as an isolated given is to abstract this entity from world, to suppress the phenomenon of world (SZ 65). That such suppression occurs reflects the tendency of world to retreat, to ensconce itself in a plenum of meaningful things and relations (SZ 75, 87).

"World" refers to the open, illuminative realm, which governs a given time and age and allows the entities to arise and be what they are. World is not exhausted by the nexus of meanings and relations; for these too are founded on the basis of the open realm which bounds and illuminates. World releases items in their determinateness and meaningfulness so that we can deal with them and have rapport with them. At the same time world per se lurks in the shadows of the background, so that what is immediately and directly manifest to us are the things-that-are and their determinate contexts rather than world per se (SZ 75, 83–85).

We tend to understand our everyday encounter with the things-that-are as having the character of an active concern. We sum up our problems in terms of the immediate given contexts around us; we view ourselves as doers, agents, who order and arrange, produce and create, improve and alter the things and situations around us (SZ 56–57; Hum 5). The idea of an agent implies a substance, permanently present—a *quid*, whose presence is nonproblematic, but whose actions are problematic. For Heidegger, however, the very presence of man is highly problematic. We are born into the world and can die at any moment; being born and dying are borderline cases, as are also waking and sleeping, which cannot unambiguously be called acts. Therefore the idea of an agent does not for Heidegger exhaust the whole of man's nature.

Man is at most a finite agent; in this case what is important is his finitude just as much as the fact that he is an agent. A finite agent is dependent upon something other than himself in order to execute his act. Thus in order to hammer we must have something like a hammer; but we must also have a nail and a board. In order to build a house we must use materials which we did not ourselves create. The ultimate conditions under which we can act are not at our disposal.

The question arises how far are we justified in viewing a finite act in terms of the unambiguous schema of agent-act? If hammering can only occur when there is a hammer-like object at our disposal, is it correct to say that we are the ones who hammer? On the other hand the hammer does not hammer by itself; man must wield and regulate it. Man is still in some sense a necessary condition for hammering. We *let* there *be* hammering, given that there is an appropriate opportunity for such hammering to take place. Our finite doings can therefore be characterized as "lettings."

Although Heidegger never explicates the phenomenon of letting in much detail, there can be little doubt that he believes this phenomenon underlies and permeates our entire rapport with the things-that-are. In *Sein und Zeit* our rapport with things-at-hand (*zuhanden*) is described as a letting acquiesce into engagement (*bewenden lassen*) (SZ 84–86). In the essay on truth Heidegger states that our most elementary rapport with the things-that-are is in essence a letting-be, an *acknowledgement* (and not a mere perception) of these things as they are within the open realm (WW 14). Finally, in the *Introduction to Metaphysics* Heidegger remarks in passing that our affinitive link (*Bezug*) with Being is a letting, and that all acts and willings are founded in this essential letting (*EiM* 16).

The word "letting" does not designate a simple process, where we can unambiguously pinpoint the role of each member of the complex involved. Indeed, the very word itself is ambiguous in meaning. For example letting may be a permitting, as when we let someone or something under our jurisdiction fulfill a wish, need or propensity. An executive may let his secretary have a coffee break, but he might let her have only one each day. Letting in this sense prescribes boundaries *within which* something or someone under the jurisdiction of the lettor is given free reign. The executive does not bind the secretary to him; he lets her have a certain amount of latitude, a certain amount of open freedom with which she can move about. He does this not out of generosity, but because he is finite and does not have absolute control over the secretary; if he abuses her, she may quit. The same is in principle true for the hammer; we cannot engage in hammering if we bind the hammer to our person. We must allow the hammer a space in which it can move freely, but at the same time we must keep hold of the tool.

Another kind of letting is a noninterference, an omission. If someone is doing something in a way we don't approve of, we often let him go ahead, perhaps for fear of offending him, or because we are no more confident of the way we would go about doing it than we are of his method. Similarly, because we are reluctant, or because we feel ourselves unqualified to decide an issue, we may let it decide itself. These examples suggest that "letting" is a withdrawal from, a compliance with, or an acquiescence into. It should be noted however that by letting something be in this sense we do not completely refrain from acting.

Letting thus has a dual sense of permission and omission. What is important about both senses of "letting" is that they involve a compliance on our part with conditions which we do not control. Letting is a kind of mediated doing, which refers to and is contingent upon something else. Thus whether we *let* George *do* it (permission) or let George do it (acquiescence), both kinds of letting involve George.

Both senses of letting acknowledge George as something over against us; both lettings let George have his own integrity, they let him be what he is. Thus both senses of letting are founded upon a *letting-be* which first acknowledges the integrity of George and allows him a free space within which to show the various aspects of himself. Letting-be allows the unconcealment of George as he is (WW 15).

It was seen earlier that world, the open realm, releases the things-that-are so that we can deal with them, arrange, order, care for and use them (SZ 83). This release is not absolute however; we are not given complete mastery of these things, but are rather allowed to operate within a bounded, open realm. We are accustomed to calling the bounds of this free realm "law"; what orients us within our bounded, free realm is our circumspective judgment (*Umsicht*), which must focus simultaneously upon the bounds and upon the specific context immediately encountered. Thus circumspective judgment is also permeated by the oscillative, reflexive relation. For example, a nail must be hammered into a board; there is no hammer in the immediate, open realm; only a feather lies nearby. Our circumspective judgment tells us that the attempt to drive the nail in with a feather is "out of bounds"; the feather was not *absolutely* delivered over to us, but was delivered over in a bounded, feather-like, open latitude which regulates the ways in which we can behave towards the feather. We are thereby forced to acknowledge the feather's featherness, the bounded realm of latitude in which the feather is delivered over to us *as* a feather-like object and not as an "all-purpose," indeterminate lump which is ready to follow our every demand and caprice.

The way in which we correspond to and acknowledge the open realm per se, is through *letting-be*. As Heidegger puts it: "Letting-be ... means letting oneself in on the open realm and its openness which each and every thing-that-is stands into, the openness, as it were, bringing that thing along with it" (WW 14).[7] Letting-be is an acknowledgement both of the item itself and of the open realm or free latitude within which this item can move and still be what it is. Letting-be is thus a two-fold rapport; (a) with the things-that-are, and (b) with the open realm within which they appear as what they are. The more essential and fundamental letting-be involved an explicit recognition of the open realm per se, such that we no longer take this realm to be the mere potency or possibility of a particular object but rather a dynamic field of latitudes: world, the open horizon of Being. To do this, we must relinquish our riveted "factical" insistence on particular things and situations; we must let these be by letting them belong to the open realm per se rather than attempting blindly to master them and to subjugate them to our will. We must "ex-ist" rather than "in-sist" (WW 21).

Only by relinquishing our insistence upon particular items—only by letting them be in the sense mentioned above—can we encounter the open realm and find a meaning (Sinn) to Being. This will prove to be a central point in Heidegger's philosophy; for if we insist solely upon the items themselves, we are insisting upon them by closing ourselves off from the open realm as such. This notion of Heidegger's strikes at the very heart of metaphysical endeavour; he maintains that the attempt to master, order and arrange the things-that-are into a closed system conceals Being and thereby sows the seeds of nihilism (WM 20; ZSf 38; VA 62, 77–78).

The question now arises, is the acknowledgment of the open realm an arbitrary act on our part, or is this acknowledgment somehow compelled? What might compel us to recognize the open realm per se, if the items therein are more immediately manifest than the realm itself?

Our essence laid claim to by something other than the things-that-are

Just as man's relation to the thing-world is not exhausted by the schema of consciousness-of-an-object, so too for Heidegger our relation to ourselves is not simply a matter of self-consciousness. Rather, Heidegger characterizes the human, reflexive relation through a complex phenomenon which he calls disclosure (*Erschlossenheit*) whereby we stand open to the world.

Disclosure does not simply involve an "I" over against a world of objects; indeed, the problem of separating off what belongs most truly to the essence of man is a highly subtle problem (SZ 321–22). For example, in the mode of everyday *Seinsverständnis* we understand and interpret the essence of human existence in terms of the thing-world (success, money, and so on) or in terms of concrete conventions and laws (SZ 22, 42–43). Consequently, the primordial character of human existence lies for the most part concealed (SZ 222, 311).

That the whole of human existence is not exhausted through our rapport with the thing-world or with the laws and conventions of human society is shown by the fact that we can become weary or rebel from these things. Nor do the same things concern us from day to day; we grow older and develop, and the things around us change. What was five years ago a concern for a seed is now a concern for a young tree. Whereas we were once concerned with toys, we are now concerned with implements. This would indicate that concern itself is not exhausted by specific objects; the objects in and of themselves do not give rise to concern. The concern must somehow be inherent in factical existence itself; concern expresses the *relation* of man to that which determines him as what he is and places him in the midst of the things-that-are. The relation of man to world is one of concern (SZ 192–93).

If man is primarily related to world per se and also (by virtue of this relation) placed in the midst of the things-within-the-world, then man's dependence upon these things cannot be the primary mark of his finitude. There must be a more fundamental conditioning structure inherent in world itself. It should be remembered at this point that world is not something which stands over or against us. We are in a certain sense congruous with world; were this not the case, we could not traverse the open realm, nor could we stand open to world per se and thereby encounter the things-within-the-world.

The way we abide has been characterized in terms of factical existence. This expression does not denote an objective, conditioned "that-ness"; rather, factical existence has the nature of a disclosure, a standing open. This standing open can either orient itself primarily in terms of the things-within-the-world or it can be explicitly reflexive and disclose its own openness (SZ 42, 188). When it does so fully, the phenomena of dread, ability-to-die and conscience make their appearance (SZ 305–8).

In *Sein und Zeit* Heidegger describes disclosure as having four basic and concurrent constituents: state-of-being (*Befindlichkeit*), understanding, speech and fallenness (*Verfallenheit*). These together comprise the ecstatic character of human existence (SZ 365).

Our state-of-being discloses the qualitative totality of all-that-is. This totality manifests itself separately from the part-whole schema and is thus more a fundamental, dynamic unity than an abstract plenum or entirety (WM 30–31). This dynamic unity is directly disclosed to us through our moods, affections and dispositions—in other words, although an attunement need not be mediated, this does not yet entitle us to write off this phenomenon as something "irrational," belonging to the realm of feeling and caprice.

Because we are attuned, emotions and moods can "come over us." Attunement thus accentuates the fact that we are conditioned by something which we do not ourselves control. This "something" is not an entity but is itself congruous with world, and therefore cannot be encountered as a discrete item. All comprehension and letting-be presuppose attunement. We cannot engage ourselves in a given activity if we are not in the mood; we will not listen to others and comprehend what they say unless we are in a basically receptive mood, unless we are so attuned as to listen (SZ 134-35).

In the mode of everydayness our attunement is for the most part neutral and indifferent. Such attunement avoids the direct disclosure of the human essence and tends to lose itself in the affairs of the thing-world. Our taking flight to the thing-world thus conceals and dissembles the original disclosive character of attunement, since we thereby interpret the causes and natures of our moods in terms of the things around us rather than as modes of disclosure itself (SZ 136-38). The fact that our involvement with the thing-world does not exhaust our essence is shown by the fact that the thing-world can slip away from us; it can lose its importance. When this happens we are brought before our own openness and thus before the open realm in all its austerity. The attunement appropriate to this bare disclosure Heidegger calls dread (SZ 186-91).

Understanding is the aspect of our disclosure whereby we are projected out beyond and beside our immediate affairs. In understanding we are able to traverse the open realm and gain many perspectives on the things and other people in whose midst we are. We can discover new things; we can push beyond a set of fixed circumstances (SZ 145-47). The ability to be beyond our immediate circumstances, to traverse the open realm, is not something which we voluntarily exercise. We are always beyond ourselves in some way because we are ecstatic; we are, as it were, necessarily free. Thus our understanding is conditioned, and therewith also attuned. The "emotionless" theoretical attitude of "pure understanding" is as much an attunement as are the intense attunements of joy, dread, sorrow (SZ 136, 144-45).

When, through attuned understanding, we traverse the open realm, we can either direct ourselves toward the things within the open realm or we can attempt to fathom the bounds of the open realm itself. These alternatives are not completely free choices, insofar as we are forced to direct our attention toward an object when it immediately and violently importunes. Even then it is not the object in and of itself which commands our attention, for an object can only importune in such a way if we are distracted away from our own openness. The tendency to forget our own openness and to distract ourselves into the thing-world is known as *fallenness*.

Fallenness is as necessary an element of our disclosure as attunement and understanding are. Attunement sums up the qualitative totality of our situation—our facticity. Understanding pertains to our ability to freely traverse the open realm. Fallenness expresses our tendency to rivet our factical existence, our attuned understanding, to the immediate thing-and-people-world, the world of everydayness (SZ 175). In fallenness we do not traverse the full bounds of the open realm; instead we rivet ourselves upon a limited aspect of this open realm, namely the actual now of the revealed items. The past and the future come to have inferior status to the present; the past comes to have only causal significance, and the present is only considered significant if it can be actualized. The standard for "reality" becomes the actual-now, be it a lowly, transient thing like an appetite or a lofty, eternal thing such as Aristotle's Unmoved Mover (SZ 346–47).

On the other hand, if in understanding that our attention is directed towards the open realm and its very openness as such, we are brought before the very limits of our ability to be "in" revealing/concealing. This limit we know as death. The significance that even the later Heidegger attaches to death is shown by his reference to it as the "shrine of Nothingness" and the "mound (*Gebirg*) of Being" (VA 177). Both the idea of a shrine and the idea of a mound suggest a togetherness of revealing/concealing. The function of a shrine is to allow something to be present while absent. A mound ensccnces something, preserving it from total oblivion by sparing it from the wear and tear of exposure that things suffer in the open realm. Graves are mounds. They are also shrines. Thus even our ontic behavior towards death seems to attest to its shrine character and its intimacy with revealing/concealing.

Death means that we have not been granted absolute presence in the open realm; in fact we are being consumed or used up by the open realm and its dynamic of revealing/concealing. We have been given a *while*, an open time span, which is bounded on all sides by absence and concealment. Yet it is this dark, empty boundedness, curiously enough, which enables us to view our life as a whole rather than as an unbounded, aesthetic, day-to-day existence (SZ 261–65, 305–6).

The open realm in which we move about and spend our days ultimately belongs to and is laid claim to by Being rather than by us. Because the realm is bounded on all sides by concealment and mystery, the bounds of this realm are eerie and forbidding. The very attempt to penetrate these boundaries seems to compel our return to the familiar, everyday world or pay the price Semele paid for trying to go beyond the bounds of the open realm in order to see Zeus as he appeared to the gods. Perhaps death as the "shrine" or the "mound" is the boundary-point itself of revealing/concealing. In *Was ist Metaphysik?* Heidegger speaks of a sacrifice or a willing departure from the things-that-are and onto the path for the maintenance of the benign bestowal (*Gunst*) of Being (WM 49–50).

An animal tries to preserve and prolong its life out of instinct. For a man, who knows that he is ultimately doomed to die, this seems futile. Many find it bizarre that huge quantities of money are spent in hospitals trying to stretch out a doomed man's coma a few hours longer. For us self-preservation and prolonging of life is always haunted by the "dark winds." Moreover, sacrifice is an act that goes back to the earliest of known civilizations. Sacrifice involves the principle that a man can choose to die rather than live; that he realizes that some things may be more important than his own being.[8] For this reason Heidegger makes a big point of the fact that only man can be open to his own dying; only man grasps his death in advance, as a possibility.

The ability to die remains the final and unmistakable mark of man's finitude and of the fact that something other than himself lays the ultimate claim to his essence, whether he is a king, a brilliant engineer, a valiant soldier or a field-worker. Hence Heidegger makes the statement: "In death human existence must simply 'take itself back'" (SZ 308).[9] Death involves self-cancellation, a radical taking-ourselves-back. We do not own the open realm; we are let reside here by something "beyond" us.

The ability to ecstatically traverse the open realm to the bounds of our existence presupposes that we can be brought out of the everyday state of fallenness. One way this happens is when dread overtakes us and we are made to feel the alienness of the things around us; we are thereby brought openly before our own congruousness with the open realm (SZ 187). Usually however we flee from this bare disclosure of openness; in fallenness we seek distraction in the thing-world, we seek the company of society, we seek the certitude of conventions and forms (SZ 184). By thus "losing" our self, we can render ourselves in large measure insusceptible to dread. Therefore what decisively hauls us out of our fallenness cannot be dread alone.

What brings us *decisively* back to our self, back to our own unique nature as (openly) being-in-the-world, is the call of conscience (SZ 269). Conscience is not merely the ability to morally censure our own activities, nor is it simply the "voice within." The call does not come from a discrete entity; it is not voiced, nor does it disclose factual information. The call comes to us in the *modus* of silence (SZ 273).

The call is possible because speech per se, of which silence is a modus, is a fourth basic constituent of disclosure in addition to attunement, understanding and fallenness (SZ 271, 160–65). Speech —that is, the ability to speak, language—articulates our attuned understanding, both in its authentic and inauthentic state, thereby allowing us to orient ourselves within the open realm (SZ 161). Only because the open realm is so articulated can we encounter particular items as the things they are, for example, houses, trees, dogs. *Thus language somehow cuts across the things-that-are and the open realm* (SZ 349–56). This "intermediary" character of language comes to explicit expression in this statement of Heidegger's found in his essay, "Wozu Dichter?": "When we go to the wellspring or through the forest, we always and already go through the word 'wellspring,' through the word 'forest,' even if we do not express these words and do not think about anything connected with language" (*Holz* 286).[10] The problem of language in the ontological difference will be taken up in detail later. For now, it should be noted that language provides the bridge whereby we can "cut across" from fallenness to authentic disclosure and vice versa. Why and how this is the case will have to be brought out later.

Conscience is a mode of disclosure which unambiguously refers to the self and sets it off from the superficialities and conventions of the thing-and-people-world. Once we are so brought before our own self, we become susceptible to dread; this in turn frees us from the things immediately around us, for these lose their importance. Thus we become free to traverse the open realm to the very bounds of our existence. The call of conscience awakens us out of our forgottenness; as such, we respond to the call to the extent that we become aware of the bounded, open time span as something belonging not to us but to something which lays claim to our essence and conditions and allows us to be what we are.[11]

The analyses of dread, death and conscience have, as many interpreters of Heidegger have pointed out, an accent on something "negative." This "negative" character is associated especially with Heidegger's earlier works, which he himself characterizes as too much bound to the language of metaphysics (WW 26, Hum 17). However, although the notion of radical finitude remains throughout the whole of Heidegger's work, the "negative" phenomena mentioned here—dread, death and conscience—would seem to have undergone a transfiguration. Indeed, such a transfiguration must be implied in the overcoming of nihilism, since what is other to what-is must lose its "negative" and "inferior" evaluation through such an overcoming. Being, the open realm, although not something-which-is, must not be lumped together with the absolute Nothing. Heidegger's doctrine of Nothingness involves a kind of dialectical argument, insofar as the encounter with Nothing is our experience and interpretation of Being when we make the things-that-are our criterion for reality (WG 5). From the standpoint of an individual who is just thrown out of his everydayness, the retreat of the thing-world is regarded as a loss, and the exposure to openness per se is viewed as an encounter with the Nothing. Such an individual experiences dread, for he still insists upon the thing-world as his criterion for what is real. However, the encounter with Nothing can only be regarded as negative and nihilistic if we assume that what is other than the things-that-are *must* be Nothing; it is just this rigid either-or which Heidegger tries to question[12] (WM 22; 2 Ni 53–54).

Authenticity

The phenomena of fear and dread (Angst) show us most clearly that there are other things different from us, which threaten us, demand us, or just leave us cold. However fear and dread are very different themselves; fear is an emotion directed toward the things-that-are, whereas dread involves the open horizon of Being itself. Fear discloses to us the hazardousness of our being in the midst of things. Fear has as its polar opposite hope, that is, anticipation of particular things-that-are-to-be. Thus, as Aristotle also observed, most fearful situations involve man in a vacillation between fear and hope. Typical of this are the man who must have surgery, the maiden looking forward to her wedding day, the graduate student who must pass his qualifying exams, the expectant parent. Dread, on the other hand, has as its object *nothing*—only sheer openness itself. In dread no particular thing threatens us, there is nothing to look forward to, only death awaits us ultimately. What comes over us is a feeling of homelessness: amidst the things-that-are, we are also other than they. Thus Heidegger speaks of a *Vereinzelung*, a kind of arrest in solitude (*SZ* 188, 298).

Fear is, then, a phenomenon of fallenness. Thus both poles of existence, the things-that-are and the open horizon of Being, involve negativity. Both can give rise to flight. In the case of fear the flight is from one thing-which-is to another, and possibly another. In the case of dread the impulse is away from empty openness to the silent thing-and-people-world, but in vain. We can attempt to flee dread; but the silent call of conscience cannot be drowned out. Once dread overtakes us, we cannot flee. We are brought up before our own doom, the Nothingness that ultimately awaits us in death.

There is a connection between dread and transcendence, in that it is through dread that transcendence is most immediately disclosed. Why is it really that we cannot drown out the silence of the call of conscience with all our portable radios and tape recorders? Or perhaps idleness is the devil's workshop; we might try immersing ourselves into our work or our families and friends so deeply and so intensely that we have no time to worry about the meaning of things. But here is where transcendence makes felt its sting; slowly the foundations of all things-that-are erode away and are replaced by new ones. Transcendence means change: reveali*ng*/conceali*ng*. Sometimes the pace is sudden and drastic; sometimes it is slow and subtle. Still, it is all the same. Implied in going-beyond is a slipping-away (*Entzug*), a withdrawal. For this reason Heidegger later uses terms such as withdrawal (*Entzug*) in place of transcendence to characterize this part of the dynamic tension of human existence (*WhD* 5–7).

Yet in spite of this a man may refuse to face up to his dread; he may allow it to become dissembled as fear. In doing so he allows his own identity to become dissembled; he becomes inauthentic. There is a connection between inauthenticity and fallenness. Both amount to preoccupation with the things-that-are while remaining oblivious to the open horizon of Being. However, inauthenticity also involves self-interpretation; fleeing from the openness of the open horizon, we attempt to find our meaning and identity in the world of things-that-are: work, family, friends. He is the doctor, the youngest son, he loves her. That is who he is. It sounds like an obituary; for an obituary is precisely a last attempt to sketch an identity where there is none, where there is only "a gap, an in-between."

Oddly enough, Heidegger does not consider the possibility of an inauthenticity based on neglect of the world of things-that-are and preoccupation with the open horizon itself. There is no problem of *Vergessenheit* in this direction. This is due to the way in which the revealing/concealing originally opened up: what was made accessible to man was the things-that-are. The opening-up of revealing/concealing did not come to light. Man continues to spend his existence in the shadow of this refusal of the opening-up to show its inner heart. Revealing/concealing withdraws always from man and draws him along in its wake (WhD 5). Sometimes he is drawn along violently and rapidly; at other times he is drawn along slowly and quietly. But here again the result is the same; no matter how fast and how far man is drawn along, it is not far enough. The open horizon remains beyond him, and he is left behind, amidst the world of things-that-are.

Thus man is a kind of homeless wanderer. And yet Heidegger speaks of man as a dweller. How can one dwell where there is no home? What sort of a being can be homeless? Only one that in essence can have a home. Stones are not homeless; metaphorically speaking, perhaps we could say that world is their home. But world is not man's home because world continually withdraws from man.

What does it mean to be homeless? It does not mean the lack of shelter. Being able to dwell and having a home are two sides of the same coin; where there is the one, there is the other. On the one hand we must have a home in order to be able to dwell; yet it is through dwelling that we make a home. To dwell is to sojourn, to spend one's days. Dwelling suggests the idea of lingering but also the transitoriness of this lingering. Dwelling has a nomadic ring to it; our wandering in the wake of the withdrawing horizon is precisely what our dwelling amounts to. Thus Heidegger is fond of quoting the line from Hölderlin: "... poetically dwells man upon the earth" (VA 187–204).[13] Apparently it is only through poetry that the terms of revealing/concealing come into light for man: earth, world, and the holy source that remains concealed. Dwelling must then be explicitly upon the earth, beneath the heavens and in the hidden face of the Holy. Man only sees himself as he truly is when he interprets himself as a member of the Foursome: man, earth, heavens, and the hidden gods which are concealed in the openness of the heavens. So abiding, man dwells.

Yet for all this, dwelling cannot be interpreted as some sort of in-

Man and Ontological Difference 63

ternal fantasy act. Heidegger is not a Romantic. Earth is real, heavens are real, the hidden source is real, man is real. Dwelling upon the earth, even poetically, involves dwelling amidst the things-that-are. We dwell by cultivating the fields, erecting edifices, producing artifacts. But how is it possible to be amidst the things-that-are without being taken in by them? Must the things-that-are and the open horizon always compete for man's attention so that he cannot acknowledge the one except by turning away from the other? Is man, the being in between, able to come to terms with his own betweenness?

In his short work entitled *Gelassenheit* Heidegger takes up this question. First he points out how man has become enslaved by technology. The freeing of man from his immediate concerns and weariness of backbreaking toil has not liberated him to be open to the heavens and the mysterious self-concealing holy source. Rather, technological man is bound all the more decisively to the things-that-are (*Gel* 14–15, 17–18). Yet this is not the fault of technology itself, nor would the situation be at all remedied by means of a blind repudiation of the technological world. Indeed, being liberated from the immediate oppressiveness of weary toil *could* free man so that he could traverse the open realm and recognize his membership with it. It is not while we are working, but while we are at rest that we have the opportunity to authentically turn our attention toward Being. While we are at work our attention must be focused upon our work; we must *lose ourselves* in the details of our work, and this means that there is no room left at that moment for directing our attention toward the open realm, Being.

The accomplishments of technology could thus be of actual aid in allowing man to turn his attention toward the open realm as such. The problem lies in the way we encounter these objects and accomplishments. As Heidegger puts it: "We can take the objects of technology into use as they must be taken. But at the same time we can let these objects acquiesce in themselves as something which does not properly concern our innermost nature. We can say 'yes' to the everyday use of the objects of technology, and at the same time we can say 'no,' insofar as we guard ourselves against them so that they do not lay exclusive claim to us and warp, confuse and finally lay to waste our essence" (*Gel* 24–25).[14] This statement shows us that Heidegger thinks it possible to have authentic encounter with the things-that-are; only we must not try blindly to master these things as if our whole significance were rooted therein. We must at once say "yes" and "no" to the thing-world; we can utilize these objects, but at the same time we must let them have their own integrity, we must *let them be*. This simultaneous affirmation and relinquishment of the items in the technological world Heidegger terms *Gelassenheit*, aquiescence in letting-be (*Gel* 25). To sum up, the nature of our authentic rapport with the things-that-are is aquiescence in letting-be.

When we relinquish our absolute "claim" on the world of entities, we allow ourselves at the same time to be open and receptive to Being. But it is Being which first allows the things-that-are to appear within the open realm. Thus letting-be is not merely a letting where we ourselves permit the things-that-are to be present; letting-be is above all a letting-Being-let-be, that is, a relinquishing of our own claim on these things and letting them belong to that by virtue of which they are granted their presence. Thus we recognize our own finitude; we recognize that "something else" is at play here which conditions our own rapport with things. It is neither us alone nor the hammer alone but Being which ultimately allows that there be hammering. In letting-be we comply with certain conditions; we allow Being to use us (VA 68–69).

In the book *Gelassenheit* Heidegger calls being-open to Being an openness for the Mystery (*Offenheit für das Geheimnis*) (*Gel* 26). By being open to the Mystery we recognize the contingency of our present scientific knowledge. We grant the possibility that a given fact "could always be otherwise." In the language of *Vom Wesen der Wahrheit* we ex-ist, that is, we stand open to the open realm of Being, rather than confining ourselves insistently by making absolute judgments about particulars (WW 21). In the openness of authentic disclosure we admit the possibility of something unknown, even contradictory, to our world; for we *put into question* our own faculties—for instance, reason, will, the senses—rather than blindly measuring and evaluating what is real on the basis of these.

In *Sein und Zeit* the analyses of dread, conscience, death and guilt show that something other than the world of entities seems to lay a claim on man's essence. Even the most distracted person is capable of being overcome by these phenomena so that he is forced to "drop the mask of superficiality." These are the primary phenomena which serve to haul man out of his state of fallenness; he may be decisively hauled out, or again the authentic disclosure may last only a split second, whereupon the individual again takes flight. It should be re-emphasized that this does not yet mean that Heidegger looks at man's authentic rapport with Being as something categorically severe, ascetic, and negative. These phenomena themselves only assume negative significance when they are viewed and evaluated in terms of criteria which pertains solely to the entities and our everyday wants. But in addition to this, once authentic rapport with Being is decisively reached, something more of the nature of a tranquillity sets in. Once we have become aware of the claim which Being makes upon us, if we persist in heeding this claim through the acquiescence in letting-be, we become less subject to the violent sting of being hauled up before the openness of Being. Perhaps this is the reason why the later works of Heidegger often reflect a tranquil mood rather than the menacing disclosure portrayed in *Sein und Zeit*.

If we respond to the call of conscience our self is disclosed in the mode of resolution (*Entschlossenheit*). Heidegger calls resolution the authentic, primordial disclosure of the phenomenon of our being-there in the open realm (SZ 297–99). It is in this mode that we, while together with things and with other people, clearly differentiate our essence from what we see in the world. Thus resolution is not an aspect of will but rather of disclosure. Heidegger equates resolution, the explicit relinquishing one's absolute claim on the things-that-are and holding oneself open to the realm of Being, with the acquiescence in letting-be (*Gelassenheit*) and the openness for the mystery or forbidding boundedness implicit within this openness (*Gel* 61).

Authentic truth (resolution, acquiescence in letting-be) is an immediate, nonmediated disclosure of Being, a disclosure which bypasses the thing-and-people-world and therefore cannot be described in categorical terms. This disclosure cannot be an emotionally indifferent kind of rapport, that is, it cannot be "theoretically objective"; for not only our rational faculties (indeed, these perhaps least of all) but also our emotions and actions—our state of attunement—bear a reference to Being and our open placement in the open realm. One of the reasons why Heidegger gives certain poets a unique status is because the poet has this highly intimate relation to objects and to Being. When the poet writes about something in the world he has adopted an attitude of letting-be towards the object itself. But more important is the fact that the poet tries to express the original disclosure in an intimate way rather than one filtered through theoretical categories; he makes no attempt to subdue certain aspects of his rapport as does a scientist.

Sometimes Heidegger contrasts authentic rapport with indifference (2 *Ni* 254). By indifference he means the darting from particular to particular, forgetting one as soon as another captivates our momentary attention. In such a state we are unwilling to hold our attention on anything except the obvious which captivates us momentarily; we make no attempt to preserve the revealedness of something (SZ 172–73). For example in a social conversation someone might reveal something vital, but no one makes an attempt to preserve the revealedness of this item. The subject of the conversation never stays the same; as soon as something is revealed it is again forgotten. When we are indifferent, we do not care whether something stays revealed or whether it sinks into oblivion.

The attempt to preserve the *revealedness* of things is an authentic type of rapport, since we then view the revealed particular as a kind of instrument whereby revealedness per se comes to light. We do not take the particular for granted but we recognize it as a gift, as something which was allowed to appear (VA 39–40). There is an old expression, "It's not the gift but the thought that counts." This expression is based upon the fact that the recognition of something as a gift takes us beyond the mere whatness of the item given. It is not totally a matter of the item itself; the emphasis is put on the item's having been given. Thus the item qua gift is an expression of something higher and more fundamental then the item's whatness.

Things are revealed in that they are given, allowed to appear. There is no agent who merely dispenses things; rather this giving is more akin to destiny or fortune. However this is a matter for later discussion. What is of interest here is that an item in its revealedness is recognized not merely as a "what" but as a gift, as an expression of Being. We are not taken in or captivated by the item itself, but rather our concern is a concern for the item qua the expression of the "it-is-given" aspect of Being. In discussing the German expression *es gibt* as in *Es gibt das Sein* Heidegger says: "The 'it,' which here gives and is given is Being itself. The 'gives' names however that which gives, the abiding presence of Being, granting and allowing its own truth to thrive. The self-giving into the open with this 'itself' is Being itself" (*Hum* 22).[15]

An attempt to merely preserve the item itself in isolation is inauthentic preservation. For example an art collector may collect and preserve valuable art objects, but he may take their essence totally for granted. He looks at the objects in terms of financial and cultural worth rather than as gifts or disclosures of Being. He does not ask himself what art is or what an art object is; he merely judges whether a specific object is "good" art or not and his judgment may be rather superficial.

In *Sein und Zeit* Heidegger mentions three modes of temporality according to which we orient ourselves in the open realm: the future as that-which-is-to-come, the present as immediate encounter, and the past as that-which-has-been. These modes may be disclosed authentically or inauthentically; in the former case the three modes are given a more nearly equal weight with no radically sharp lines drawn in the sense of "now," "gone by," or "not yet," which would give the present mode predominance over the other two.

The authentic future is a traversing the pure openness of the open realm, holding ourselves open for the Mystery; the inauthentic future is an anticipation of particulars. In other words the authentic future is disclosed to us such that we gain a synoptic vision of the openness of the open realm per se; the inauthentic future is only disclosed from the perspective of the present (SZ 337).

The authentic present Heidegger calls *Augenblick* or moment in contrast to the inauthentic present of nowness. The instant spoken of here is not a "now" which is cut off from past or future. It is a decisive moment in which we receive insight into our *situation* as a totality: not only do we live in a thing-and-people-world, but we are ecstatic beings, congruous with the open realm that we are traversing. The moment is in a sense a twofold openness on our part, an attentiveness or readiness to encounter something in its revealedness, while on the other hand guarding this openness or receptiveness by restraining ourselves from being totally captivated by the particular items we encounter (SZ 338).

The authentic past places the emphasis on what-has-been rather than on what has gone-by. The former still lingers with us and continues to be operative; what-has-been is the greatness of tradition, for example the metaphysical tradition. Something gone-by is a petty quarrel or one's childhood. Heidegger describes the disclosure of the authentic past as a recapitulation (*Wiederholung*) in which what-has-been is brought-back (*wiedergeholt*). The disclosure of the inauthentic past is characterized as a forgetting. Forgetting is a kind of concealing—indeed it is the kind of everyday oblivion in which we find ourselves *zunächst und zumeist* (SZ 339). The ability to forget this or that is rooted in a deeper ontological forgetting which occurs as the concealing of the open realm itself (SZ 219; VA 264). Because of this deeper forgetting, all things in the open realm are subject to being forgotten. The attempt to recall this or that is an inauthentic disclosure—still a kind of forgetting, because the real, ontological concealing is still forgotten. It is like the man who remembers he left his hat on the radiator but forgets that he put it there to remind him to stop at the cleaners'. His act of remembering simply points up the more profound and original forgetting. When we forget the open realm, the things-that-are do not appear to be assembled in any kind of common presence. Hence we concern ourselves with those objects that captivate and concern us most immediately and most intensely; the others have simply gone by, fading into nothingness. This superficial remembering of this and that does not preserve the past as a distinct mode of disclosure. This and that appear as weak, fleeting images compared to the now-actual; the result is that this superficial sort of remembering closes off our vision of what-has-been and rivets us to the now-actual. True recapitulation of what-has-been would on the other hand widen our sense of participation in the open realm. Such recapitulation is not a blind attempt to recapture the moment but is rather the reassemblage of all revealed things into a common presence so that we may see them in their revealedness. For it is their revealedness which has truly been.

In *Was ist Metaphysik?* we are told that the more authentic and essential our thinking becomes qua thinking (*je denkender das Denken wird* . . .) the more our thinking concerns itself with what-has-been thought (WM 13). There are two reasons for this: (1) By turning to the thoughts of previous thinkers we commemoratively contemplate the historical course of the disclosure of Being itself rather than viewing Being as something independent of time and hence of the course of our human existence. (2) Whatever ideas we have of Being can never be severed totally from the past but must rather incorporate and draw on these writings which preserve, albeit in obscure form, the problem of Being.

In characterizing his methodology, Heidegger often speaks of a leap (SvGr 106, 133–34, 157; ID 24–25, 32–34). What he means by this is a leap from our limited, inauthentic way of understanding things to an authentic way. Rather than understanding everything from the standpoint of the "now," that is, where the past is "forever gone by" and hence to be forgotten, and where the future has as its import the "not-yet," we are to leap out of this orientation into one which gives us a more synoptic view of a presence common to all three of these modes. In the new orientation we encounter the things in their revealedness while maintaining the integrity of our own disclosed position. We are openly placed before the open realm while also in the midst of things. Because we do not take the things-that-are as the final, absolute reality, we can hold ourselves open for the Mystery, that is, to that which, although revealing and illuminating the things-that-are, remains in itself concealed.

The authentic man does not write something off as trivial and meaningless merely because its essence is concealed; rather he waits patiently, ready to respond when the mysterious realm makes a demand. He keeps his evaluations, opinions, and so on, *tentative* rather than insisting upon them as final facts or truths. He attempts by means of commemorative contemplation to traverse the paths which other thinkers have gone; but here again he does not take these paths as absolutes or even as givens, but rather as gifts. When man is authentically responding to the demand which the open realm makes upon the human essence, he does not stay with what a thinker explicitly thought but rather tries to determine, in light of the historical course of the open realm, what has been left unthought (*ID* 46). Because this is so we need not engage in mere repetition, but rather we go back further each time in an attempt to reach the primordial phenomenon, the *phainesthai* (self-showing), whereby all things arise. In recollection we thus attempt to complete the authentic reflexive relation, we attempt to let-be-seen self-showing, thereby acquiescing into the fundamental unity between *phainesthai* and *logos*.

Thinking

Throughout the history of philosophy thinking has been considered the identifying mark which singles man out as unique among the earthly things-that-are. Man has been characterized by the metaphysical tradition as the rational animal, the animal that calculates and thinks. Presumably, thinking is not some sort of accidental or inconsequential attribute like having an appendix or being fond of bananas. Thinking has been valued by many as an excellent activity, perhaps the highest activity. In this latter sense of the term, animals and machines do not think. What, then, is thinking?

One of the prevailing conceptions of thinking which arose with the modern period of philosophy and continues into the present day is that of some sort of mental act that "represents" objects (*das vorstellende Denken*). There are various theories as to how this act is executed, but for the most part they all make the assumption that the mind is the sole actor and that what it represents is lifeless and passive in itself. Moreover, each act of representing, considered as an act, is formally identical to every other such act. Thus the act of representing is indifferent to the content or object represented. A thought about what one will eat for dinner is, according to this theory, no less a thought than a thought about the ontological difference. The act executed in both cases may be the same kind. Some versions of this theory allow that the thought of a possibility may be a different kind of act of representing than the thought of an actuality or an "ought-to-be," and so on. Still, within the modal field of possibility, the act of representing a possible late dinner is formally identical with the act of contemplating the possibility of the world-night. In such a view the content of thinking is not acknowledged as having a bearing on the quality of that thought as thought.

Heidegger points out that as we represent for example a tree, we can only be "successful" in our attempt to the extent that the tree presents itself before us. Hume, Kant, and others who give a formal account of thinking do acknowledge that some things are easier for us to represent than others, for example, mathematical objects are easier than matters pertaining to the realm of feeling. Yet they do not draw the consequence that this suggests—that there must be some sort of cooperative attunement between man and Being. Rather, they simply consider it a "waste of time" to try to represent clearly and distinctly what is unrepresentable.[16]

Heidegger argues that the apparent failures and withdrawals in thinking are as instructive, perhaps even more so, as the apparent successes. If thinking were solely a matter of an autonomous mind executing a formal act of representing, there would be no such failures. Thinking is not a pure, autonomous act; there are conditions above and beyond it which sometimes make possible fecund thinking, other times barren imposters of ideas. Thinking is a letting-be which moves in an element that elicits it forth. To the extent that we think, why do we do it? Why do we represent ourselves a tree? Is such thinking automatic? Some say it is a response to our environment. But this answer is opaque in many ways. Surely there are other ways we respond to our environment, for example, sleeping, working, poetizing, desiring, praying, problem-solving, even suicide.

Heidegger begins his book *Was heisst Denken?* by claiming that the matter most provocative of thought (*das Bedenklichste*) is that we do *not yet think*. He makes this claim in an age when science and academic learning appear to be flourishing as never before. The implication is that thinking is not accomplished in the various activities of science, technology, and letters. Why not? Heidegger argues that the "why" or sense of these activities, the primordial element which calls forth any thinking at all, remains concealed. True thinking must be openly attuned and directed toward its source (WhD 10). If that source continually withdraws from us, then we are always only underway. Hence our attempts to think are permeated by a "not yet."

It is most important to grasp the difference between not thinking and not yet thinking (WhD 12, 59). Animals and machines do not think. We do not yet think. Our position is one of being between, underway in thought. Heidegger likens our situation to that of someone learning, say, swimming (WhD 9). While one is learning to swim one is swimming but not yet swimming. Learning to swim means to be underway in swimming. If we are underway, we have not yet openly come to terms with our element. In the case of thinking the element withdraws, yet in such a way, according to Heidegger, that we are drawn along (WhD 5, 52).

What Heidegger has in mind is very much exemplified in the academic world today. He claims that science, problem-solving, Geisteswissenschaft, and so on, do not examine the primordial element from which they arose (WhD 56–59). Rather they opaquely presuppose this element and then proceed to investigate and establish rather specific issues within this original element. But when someone asks, why go to all that trouble—why for example be so concerned with the differences in style between Blake and Milton—the answers are awkward and uneasy. We say for example that a love of learning is self-justifying. Yet we know that what it takes to support a research project on Milton could feed a number of starving people. Academicians have to come up with various rationalizations as to why they are useful, why they should be able to command resources even if it means that others starve. They say they are preserving values, that intelligence must be cultivated in order that man will continue the long task of mastering his environment, and so on. Yet none of the answers seems wholly satisfying. It would seem that the source, the justification, of these activities is indeed concealed. If it continues to remain concealed, the academic, institutional, expensive ways of carrying on these activities may be in serious jeopardy.

It is a mistake to look for a why or sense that would be apparent from within the ranks of the things-that-are. Thinking—whether it be thinking that is transparently open to the primordial element which elicits it forth or the derivative, one-sided thinking characteristic of the positive academic disciplines, which opaquely takes the primordial element for granted—has a why which belongs to the horizon of Being. Thinking is thus our response to our environmental situation to be sure. But we must remember that man's situation in- involves the duality of the ontological difference: while being affini- tively linked to Being, man has rapport with the things-that-are.

But if the primordial element which elicits thinking forth con- tinually withdraws, if thinking is always a not-yet-thinking, what is there for us to do? To begin with, Heidegger claims that we must come to understand this withdrawal. Again the analogy to swimming will prove helpful, since we cannot master swimming by merely flagel- lating wildly in the element, water. We must occupy ourselves with what has been accomplished in swimming—established strokes, estab- lished "failures," and so on. So too in the case of thinking, we cannot think in a vacuum. Part of our element is a tradition of others who have been underway and who have left traces.

Thus thinking not only involves a primordial element but also a tradition of what has been thought and what has been left unthought. Thinking involves memory, both individual and historical. What is it to be thoughtless? A tree or a stone cannot be thoughtless. To be thoughtless is to fail to keep or have in mind. Thoughtless people do not remember. To be thoughtful is to remember, to keep in mind. Yet here again memory is not to be thought of as a formal ability rooted only in a mind. We cannot remember that which is not given to be remembered, that which is not present in the sense of being accessible for remembering. Thus memory is not merely directed to- wards the past: "Memory means originally so much as devotion: in- cessant, assembled remaining with . . . and to be sure not only with what is bygone, but in the same manner with what is present now and what can come. The bygone, the present now and that which is to come appear or come to light in the unity of presence which remains at any time singular and unique" (WhD 92).[17]

Yet memory too is permeated by revealing/concealing; this is the import of the mythical account of the origin of memory as the daugh- ter of heaven and earth. Earth, as we have seen, is present in the dual sense of *aufgehen*, upsurgence that also conceals in that it reveals, and this corresponds to the duality in presence itself. This topic, however, is not the focus of our present task, which is to see in how far thinking designates the essence of man. Thinking in its deepest sense involves memory and commemorating: staying with something. Thinking as Heidegger conceives it is contemplative; but whereas the usual ac- counts of contemplation relegate it to the abstract and mental, Heidegger emphasizes the ecstatic character of thought. When we think of something we are literally in its proximity and it is in our proximity. In the dialogue of *Gelassenheit* Heidegger claims towards the end of the dialogue that thinking in essence means coming into the vicinity of something (*Gel* 70–72). Thus thinking is bound up with the polarities of nearness and remoteness, again suggesting that thought is ecstatic in character.

Thinking involves being underway. Heidegger conceives of thinking as a kind of journey, a sojourning. Being underway, we have not yet arrived; we are not yet in the proximity of the source which elicits thinking forth. Is the problem that we are too close to the things-that-are? How close are we really to the things-that-are? The things most immediately accessible in our everyday activities and encounters are rarely things really close to us. Closeness is not a matter of spatial or mathematical adjacency. In a city two families may occupy adjacent houses and apartments and yet not really be neighbors. Thus, although we are in the midst of the things-that-are, we are not really on intimate terms with them. For this reason, the problem of coming into the vicinity of the primordial element which elicits forth thinking cannot be conceived along Platonic lines. Leaving the thing-world does not guarantee an entrance into the element of thought. In fact such an attempt does not face up to the dual situation of man: that in the midst of things-that-are he bears an affinitive link to Being. Since Being is always Being *of* the things-that-are, we cannot achieve the kind of closeness or intimacy that characterizes true thinking by simply turning away from the things-that-are.

This may seem surprising, for in the case of most differences we pursue one term by turning away from the other. In fact most differences demand this sort of behavior; the more different things are from one another, the more individual attention they seem to need. It is this kind of conception of differences that lies behind the exclusive either/or choice model. Yet because Being is always the Being *of* things-that-are, this cannot be the kind of situation Heidegger has in mind. We do not turn toward Being by turning away from the things-that-are. Thus man must achieve above all a closeness with the ontological difference as such if he is to achieve the kind of closeness which characterizes true thinking.

Yet how can we ever achieve closeness with what is continually withdrawing? Just as Being involves the duality of revealing/concealing, so too does closeness involve remoteness. In fact Heidegger claims that it is closeness itself that remains remote. Once again he underscores the theme that we are underway, that our thinking is a not yet thinking. The fact that we are drawn along means that the primordial element is not merely remote. Rather, we are caught up in a tensive interplay. Heidegger often makes reference to motions such as described by dance figures and various forms of play and interplay (ID 30, 68; WhD 84; SvGr 185–88; VA 178–79). Revealing/concealing is a kind of dance between and around and beyond us, catching us up in its rhythms.

To see why Heidegger views closeness and remoteness as an interplay rather than an opposition we need only reflect a little about what true closeness would mean. St. Francis of Assisi was supposedly close to birds and animals after he gave his riches away. Closeness does not involve ability to manipulate; St. Francis could use his money for all sorts of influence. His money would always do what he wanted. If anything, closeness involves an abandonment of interest in manipulating things. St. Francis achieved a closeness supposedly which no bird watcher can match. He did not try to know and describe and classify the birds and animals; he *let* them be. He allowed a distance between him and them, and this very distance is what made closeness possible. Only that which has distance can come close; conversely, only that which has been close can withdraw.

The interplay of remoteness and closeness characterizes the way the things-that-are are delivered over to us in their Being (Gel 68). The open realm allows us a certain amount of free play between the poles of proximity and remoteness, but this realm of free play is clearly bounded.

Congruous with the open realm and its withdrawal, we can traverse between the bounds of pure proximity and pure remoteness, we are also in the midst of the things-that-are. We can encounter these things because they are held in the *tension* of proximity/remoteness; as such they are over against us, nearby and yet to a certain degree removed (Gel 42–44). This tension is expressed in ontological terms as "the-things-that-are-in-their-Being": the ontological difference (ID 62). This tension, the ontological difference, is what makes possible and incites our thinking; for thinking is the way we correspond to our affinitive linkage with Being (Seinsbezug). And, since the latter involves us in rapport with the entities in the tension of proximity and remoteness, it is to the Difference we correspond to most primordially. The Difference, laying claim to our essence, is the primordial injunction (anfänglicher Geheiss) which summons us to think (WhD 149).

Thinking and language

Human thought is structured; we say that it is discursive. Thinking is thus not a static, passive apprehension; in thinking we must traverse a path. This path need not be linear, but it must in some way be structured, it must be traversable. The structure need not be strictly "logical"; yet it must be discrete enough so that we know we have traversed a path, that we made a beginning and that we are underway toward the vicinity of the topic of our thought. Thinking is methodical; it follows a path.

Because thinking is discursive and methodical, we say that it is articulate; in thinking we use language. Articulation allows the open realm to take on the definiteness of a path, so that we may orient ourselves while traversing the open realm (*USp* 256–57). Thus our languages are syntactical; but the syntactical character of language is not limited to sentence structure and word order. The syntactical character of language also includes the ordering of sentences into paragraphs, paragraphs within a larger schema, for example, a chapter.

The stress Heidegger lays upon language as that which blazes a path for us to traverse through the open realm separates him off clearly from the mystics. For the latter, heeding our congruousness with the open realm has the character of a nondiscursive, immediate apprehension. For Heidegger we must traverse this realm; this traversal is highly articulate and intelligible, although its intelligibility is alien to the everyday understanding. But the fact that something is difficult to understand, that it cannot be expressed "clearly and concisely," does not yet entitle us to relegate it to the realm of mysticism, where all discursive thought comes to a halt.

Because language blazes paths through the open realm, language is in a sense the bounded form wherein the reflexive oscillation of factical existence must take place (*ID* 30). Language defines the horizon of our traversal, the bounds within which we must remain as long as we dwell. In this sense, language "opens" the open realm. Thus, says Heidegger, language is the house of Being, that in this house dwells man (*Hum* 5).

To dwell authentically, according to Heidegger, is to heed, care for and preserve that in whose proximity we reside. Thereby we execute the letting-be-seen, the fundamental need of which lays claim to our essence. Letting-be-seen is thus interwoven with language; for it is only through man's traversal of the open realm that this letting-be-seen can occur. This means that language is not an instrument which we arbitrarily use; rather, language uses us in order that the traversal of the open realm, and therewith letting-be-seen, may occur (USp 259–60).

It should be stressed that, although language bounds and limits the open horizon of our thinking, this does not mean that language is a rigid structure, so that what cannot be said in terms of "accepted convention," be it logical or grammatical, should not be said at all. Language is not the totality of all existing words, nor is it a given, rigid grammatical and syntactical schema into which all words and meanings must fit and comply. Rather, Heidegger considers language as a kind of moving force which first allows words and syntactical structures to arise (USp 214–15, 261). Language has no discrete boundary, just as the sky has no discrete boundary, although it is nevertheless bounded. We do not reach the "end" of language, for we are always within its boundaries. Thus we can and should continually fathom the depths of language, finding new ways to express things, blazing new paths through the open realm.

We are accustomed to viewing our relation to language as one of speaking. When we speak we articulate our thinking. Yet, thinking was characterized as a heeding rather than a speaking. How are these two notions to be reconciled? To begin with, heeding is a decisive taking-into-regard, a responding-to. The three fundamental modes of taking-into-regard are seeing, hearing, and touch. These are not phenomenal, bodily sensations for Heidegger; he has incorporated these in ontological terms into the framework of his philosophy. For now I will consider only the two modes of seeing and hearing; touch is a much more difficult topic and one for which a more radical background must be laid.

The objects of both sight and hearing are for Heidegger highly organized. We do not first hear sounds and then synthesize these into words; we hear meaningful words (SZ 164). We distinguish between a foreign language, that is, unintelligible words, and the scraping noise of a pipe. Likewise, we do not first see colorpatches and organize these into objects; instead we see trees, horses, hammers already belonging to a significant context (SZ 67–72, 75, 83–88).

Seeing is the apprehension of that which lies before us, that is, a landscape, a single object, a picture. Seeing is our "scientific" mode of apprehension; through seeing we can best pin down spatiotemporal locations and other phenomenal aspects needed to determine and theoretically define things-that-are. We also see explicitly how an object is related to the context of things around it. Yet a momentary seeing is limited to that which lies before us; we can only see that toward which we have turned. On the other hand, that which we hear need not lie explicitly before us; it can be in our proximity and yet remain concealed. Furthermore, hearing need not give us any ideas about definiteness of an object; for example we hear the wind, we hear thunder. Since we need not be explicitly turned toward that which we hear, hearing can have the effect of bringing us before something which we have hitherto not noticed. If we hear footsteps behind us, we may turn around to see who it is. Thus hearing has brought something before us which was hitherto concealed.

It should be remarked however that seeing and hearing are interwoven with each other, and both are occurrent within the bounds of language. When we see an object it is already highly defined; we recognize it as this or that, we name it, we are concerned with *what* it is. On the other hand hearing and speaking involve seeing, not only in the case of their preservation through writing and reading, but also in imagery, diagrams, schematizations and outlines. True hearing is a comprehending; when we have comprehended we say that we have seen, we have grasped *what* the situation is.

In *Sein und Zeit* a distinction is made between discourse (*Rede*) and idle talk (*Gerede*) (SZ 169). In colloquial terms we could characterize the former as "thoughtful" and the latter as "thoughtless." A person who has the habit of weighing each word he says is usually somewhat reserved in his speaking; he is somewhat taciturn in society and not given to bombastics. Such a person usually does not speak out of caprice but rather lets "the facts" guide his speech; he is not vitally interested in letting his own personality come out. At the other end of the scale is the person who "talks to hear his head rattle." By this expression we mean that the person does not listen to the words he is saying but rather to the noise that comes out. His speech is "thoughtless," although it might be very entertaining.

The fact that speech can be "thoughtless" does not mean that speaking and thinking are independent; on the contrary, it means just the opposite. It is only when two things are interdependent that we can really speak of a privation such as "thoughtless." A stone is not "thoughtless," because thinking or not-thinking is not something which can be appropriately attributed to a stone.

The thoughtful speaker pays attention to the words that "come to him" before he actually says them out loud. He heeds them, he listens to them, restraining himself from idle talk (*USp* 32-33). Speaking is only possible on the basis of a having-heard. Even the thoughtless speaker half hears his words, although his hearing is not a heeding. He only heeds the noise that comes out, and this cannot be properly called a heeding.

It is because we are able to hear what is not momentarily the object of our attention that the call of conscience can haul us out of our state of fallenness and make us aware of the claim made upon our essence. Authentic heeding is rooted in hearing; for only if we hear the demand made by the mysterious realm will we take regard of this realm (SZ 279). The condition that letting-be-seen culminate in a seeing is that there first be a hearing of the call. Hearing as heeding is a holding oneself open and receptive to that which has not yet been disclosed (SZ 163).

In heeding we openly traverse the open realm; we think. In heeding we respond to the call; but response is speech. Mortals speak in that they respond by heeding the primordial injunction, the demand made upon their essence that letting-be-seen occur (USp 32). The source of the primordial injunction was said to be the Difference, whereby we are placed in the midst of the things-that-are, and yet our essence is laid claim to by the mysterious realm.

The Difference, far from being a mere relation or distinction, now appears as some kind of dynamic principle which accounts for and makes possible our very existence in traversing the open realm and therein encountering entities. The Difference has some kind of reality of its own; it originates the call, the primordial injunction. It makes possible the traversal of the open realm, and possibly grants the open realm itself. The problem now is to see how Heidegger characterizes the Difference as a reality in itself.

IV

Destiny and the Ontological Difference

Der Anfang verbirgt sich im Beginn.
(WhD 98)

Heidegger does not consider the Difference to be timeless and eternal in the sense that a metaphysician might say that certain truths have always prevailed, even though man had no awareness of them. Yet neither does he claim that the Difference originates in human thought as does, for example, the difference between semantics and syntactics. There must have been a breakthrough (*Aufbrechen*) or arrival (*Ankunft, An-wesen*) of the Difference (WG 15; 2 Ni 213–23). Were this not the case, the Difference would be a mere abstract relation and would have a bearing only on the abstract, theoretical parts of philosophy. The fact that each and every thing-which-is, no matter how great or how small, is a revealing/concealing of what-is in its Being (*Seiendes in seinem Sein*) would be a fact only for abstract thought and would have no bearing on the way things really are.

But when might this breakthrough have "occurred"? The arguments which show that a first occurrence is a contradictory idea come here to mind. There cannot be a "time" before the first time. Yet, this argument cuts both ways, for it can also be used to show that an oak tree had no "beginning." No origin can be pinpointed in time; conversely, any origin posits an absolute beginning in time. Hence origin and beginning are not of the same genre (WhD 150). These problems only show that the nature of origins is enshrouded in mystery. Thought recoils upon itself when we try to think these things out in the mode of ontic discourse.

If we attempt to think backward in time, we eventually reach a point where everything becomes enclothed first in myth, then in mystery, and finally in total darkness (WhD 94). Man, his world and all-that-is apparently emerged out of something; yet the emergence from this primordial origin is revealing/concealing *par excellence*. The origin itself fades into concealment as the things which it yields forth —human existence, the world, the many things-which-are—make their appearance and begin the struggle for continued presence, until they are reclaimed by that dark origin which evidently gave them up (*Holz* 309–12; 2 *Ni* 487–90).

The attempt to think backward in terms of causal explanation enjoys a similar destination. Lightning strikes suddenly from out of a dark cloud and destroys a power substation. Of course the occurrence of the lightning is "grounded"; there are scientific, causal explanations for the phenomenon of lightning. But why did the lightning strike where it did and when it did? Again, there are scientific explanations that tell us that a certain set of conditions prevailed at place P and time T. But why did these particular conditions prevail? Because others prevailed? Why? Finally the lightning flash appears demonic. We are left with our fears and our myths. The lightning is eerie and alien. It "never strikes twice in the same place." This alienness is underscored by its instantaneous character. It reveals itself through concealing itself. The dark cloud is as before. Humans are huddled around the candle light, quieting the frightened children. No one feels at ease. The ultimate *Why* cannot be answered, and yet it disquiets us. The problem of the primordial origin lurks within the lightning flash; it lurks within the occurrence of every phenomenon, no matter how spectacular or unobtrusive it might seem.

The inner unity of *phainesthai* and *logos* suggests that those who live or lived closer to the primordial origin might have or have had a truer experience of that origin. This would make the attempt at some kind of reduction meaningful, where one tries to reconstruct a relatively pure experience of something from the testimony of those who were attentively close by. This is partly what Heidegger has in mind when he turns to the pre-Socratic thinkers. However, the problem is complicated by the fact that the inner unity of *phainesthai* and *logos* also is subject to the revealing/concealing. The purer the point of origin of this Difference, the more human thought is enshrouded in myth. The closeness to this origin is self-concealing (*Holz* 310–11). Thus the hope of turning to the pre-Socratics for a final formulation of the Difference is undercut.

Nevertheless, an inquiry into the experience of thinkers through the ages with the Difference is necessary. For one thing, if this Difference is so important as Heidegger claims, then it must lie at the basis of the entire human destiny; one would expect that all philosophical thought through the ages would reflect this fact, however indirectly. But even more important is the fact that Heidegger has used a word from a language which has also emerged to designate this primordial principle: "Difference." When we name a "Difference" we commit ourselves to many things which this name connotes. What did the ancients call the break-through of man, world and thing out of the dark origin? Is the Difference ultimately a Difference?

The twofold principle as reflected in Heidegger's interpretations of the pre-Socratic thinkers

The problem of the "correctness" or "incorrectness" of Heidegger's interpretations of the pre-Socratics cannot be entered into here. Suffice it to say that Heidegger himself does not view these writings as expositions of "what the Greeks meant." Any excursus into the meaning of a text must be interpreted with reference to a framework in light of which the text takes on its meaning. The weakness of much criticism and interpretation is that it does not make its framework explicit. The phrase "what the Greeks really meant" is utterly meaningless unless we construe it within a context, for instance, anthropological, psychological, theological. Heidegger's question is, what light do these writings shed on the ontological difference? Do they grapple with the Difference at all? In what way do they obscure the Difference?

The essay entitled "Der Spruch des Anaximander" is the longest single essay Heidegger devoted to pre-Socratic thought. The essay differs from the others in that it is framed by a portrayal of the gulf between modern man and the time of Anaximander, whereas the others begin directly with their subject matter. It is this essay which lays down the methodology and framework wherein a reflective encounter with pre-Socratic thought can take place.

An immediate problem which arises is, for example: was Anaximander just one man? Or is he expressive of a whole Greek orientation? Clearly the importance of his fragment hinges on this point. Yet a "Greek orientation" also is not won in terms of numbers, for it is a kind of coincidence whether Anaximander expressed a majority or a minority outlook. "Ancient Greek" is some kind of category which requires a special approach. Somehow it stands at the outset of a long unfolding course of Western thought. It is the nature of this unfolding course which must determine our relation to "Greek" thought. Consequently Heidegger frames his essay by reflections on the nature of this unfolding course. Indeed, it is here that Heidegger formulates the kernel of one of his most difficult ideas: the wandering course of the open realm itself.

Kant had envisioned the transcendental "realm," if one may speak that way, as ahistorical, to say the least. His transcendental structures do not develop, and he evidently thought of them as binding on Anaximander in the same way as on Newton. The primitive character of pre-Socratic thought would then be explained in terms of a failure to appreciate and recognize what these transcendental principles are and what their status is. Yet he failed to explain in what way principles could be operative on thought that did not notice their presence. And if they are not operative on such thought, do they have the transcendental status Kant claimed for them? Or are they merely "nice to have and follow"?

Hegel was sensitive to this problem, and Heidegger claims that he is the only western thinker who has had a reflective encounter with the historical course of Western thought itself (*Holz* 298). Yet, Hegel has nothing to say about the fragment of Anaximander, which lies at the beginning of the long history. Perhaps the reason for this is that Hegel seems more interested in the goals of thought than in its origin, although, to be sure, the goals can only be discerned through retrospective reflection. The unfolding course thus seems to be a "stage" which will be taken up when thought reaches its absolute goal. But Hegel again seems to have no adequate philosophical interpretation of what it is to be "underway." He never fully explained why an Absolute would "need" to come to terms with itself. His philosophy is an amazing example of a position that is sensitive to the need for a transcendental framework that is epochal and dynamic; yet the fact of this need itself has no clear status within his thought. Nor can it as long as a philosophy is goal-oriented and ignores origins. In this sense a phenomenological program offers more of a hope than does a program of dialectical development.

Thus Heidegger is possibly the first Western thinker to philosophically pose the question: what does this ontological alienation, this gulf between the *Seinsverständnis* of modern man and that of the ancient Greek mean? What is the significance of the errant course of philosophy itself? Man's traversals of the open realm are not ahistorical and free-floating; they have fateful implications, for they influence the way subsequent men will make traversals; and they themselves are predicated upon previous traversals, which often become more obscure, the older they are.

The open realm itself has a kind of "drift" to it. No traversal can be exactly repeated. This drift corresponds to the drift of history itself. It is a symptom of the transcendent withdrawal implicit in the character of Being itself. If the open realm had no drift, everything within it would be encounterable for an indefinite, repeatable number of times. But that would mean there is no concealment in any real sense. The open realm would be like a timeless, sealed container from which nothing ever escapes into concealment. We would have the world of Being of Parmenides and nothing else. All things would be permeated with the light of Being. There would be no ontological Difference, no gap between the things-which-are and their Being.

But a different set of circumstances prevails. In the essay on Anaximander, Heidegger tells us: "Being withdraws itself by disensconcing itself in the things-which-are. Thereby Being leads the things-which-are astray, impregnates them with errancy. . . . Without errancy there would be no relation between the fate of one age and that of another; there would be no history" (*Holz* 310–11).[1] The "drift" then is not merely a function of human thought. This is why any interpretation of "original sin" as the source of alienation—or any other kind of interpretation that lays the difficulty at the feet of man and man alone—will be superficial. The drift is not located in *phainesthai* on the one hand nor in *logos* on the other but rather in the inner unity of both. The drift is rooted in the nature of Being itself, in the ontological difference: the fact that Being withdraws itself by ensconcing itself in the things-that-are. The open realm is revealing/concealing.

This of course complicates the approach to Being; not only is Being different from the things-which-are, but this difference is a drifting one. Being has a story, a history connected with it—the story of its hiddenness. There is no simple, timeless *Seinsverständnis* to serve as a norm for all men. Yet all true thinkers have, each in his own way, understood the things-revealed in light of their Being, overlooking the fact that the inner nature of Being is thereby concealed. The drift makes impossible an abstract approach to this problem, but it must be in some sense dealt with if we are ever to come close to the focal point of the revealing/concealing open realm in its purity. This can only be done by juxtaposing ourselves to certain points along the drift and attempting to understand what their relation is to it and to us. Anaximander stands somewhere at its beginning, as one who experienced the twofold principle which gives rise to the drift.

Why does Heidegger assume that this dawn begins with the Greeks rather than with previous civilizations such as the Egyptians or Persians? Is the dawn of culture "earlier" and more fundamental than the breakthrough of the ontological difference? Is this a defect in Heidegger's thought? It must be remembered that Heidegger is not doing history for its own sake. He is not interested in the ontic emergence of civilization as such but rather in the problem of Being:

> We seek what is Greek neither for the sake of Greece and all its glory nor in order to improve our knowledge of Greek culture; not even simply for the sake of improving our dialogue with this culture, but solely in reference to that which might be brought to word through such a dialogue, provided that it can of its own accord come to expression. That is namely that self-same which concerns cosmically and fatefully both the Greeks and us, albeit in different ways.... Greek is the early hour of the cosmic fate in whose guise Being itself projects itself as an illuminative clearing into the things-that-are and lays claim to an abiding essence of man [*Holz* 310].[2]

The rise of Greek thought is predicated on a breakthrough of the Difference, an early experience of the illuminative clearing and the concealing implicit therein. This is Heidegger's primary interest.

To Chreon: the ontological difference as brookage

The oldest known statement concerning a fundamental, twofold ontological principle is the fragment of Anaximander. This fragment has according to Heidegger been vastly misrendered through the use of physical, legal, and moral concepts. But Heidegger's contention is that the fragment is more fundamental than these technical divisions and therefore must be translated in terms that do not reek of technical disciplines. Confined to the directly quoted portion of the fragment, the typical translation reads: "according to Necessity; for they pay penalty and retribution to one another for their injustice" (*Holz* 314).[3] Moreover, writers on Anaximander generally understand this to be dealing with the physical world, and then they claim that Anaximander "applied" legal and moral concepts to physical nature. Heidegger believes that this type of interpretation fails to meet the fragment on its own ground. It sees the fragment through our eyes. No reflective dialogue occurs; the gulf created by the drift is as wide as ever.

Heidegger attempts to translate the fragment by avoiding the language peculiar to legal, physical, moral disciplines. He offers this as a translation: ". . . in accordance with brookage; for they let juncture belong and therewith also bearing-affinity one to another (in overcoming) the chaos of disjuncture" (*Holz* 342).[4]

The *sie* or *they* referred to above apparently refers to the things-that-are. Commentators have always claimed that Anaximander spoke of coming-to-be and passing away. What comes to be and passes away is what-is-occurrently-present, the entities, the things-that-are (*Holz* 315). Thus the things-that-are, those occurrently present, do not lie in a separate realm between what is about to come and what has just departed; what-is-occurrently-present has come to be, that is, has become present (*Holz* 322). Nor are things-absent totally cut off from the occurrent present; for what is absent is either that which is just arriving or that which is just taking leave (*Holz* 320). Absence does not indicate a radical privation of being but rather a remoteness of presence; there is no radical schism between the absent and the present. Instead, juncture prevails (*Holz* 327).

For example, when a certain entity "passes away," we say in common sense terms that the image or memory of it may linger on. We treat these as mere ideas which no longer refer to anything extant. Yet, if we want to talk of ideas, the object occurrently present is also an idea. In all cases we must ask, what is the status of the object from which the idea gets its meaning? The object of a memory is no less an object than the object of a perception. If a memory lingers, then there is a sense in which the object also lingers, although in a different way than when it was occurrently present. Coming-onto-the-scene and taking-leave are not sudden, black and white affairs. The occurrent present is not severed from the absent but is actually conjoined to it, shading off into absence in two directions (*Holz* 327).

All that persists or remains must reside between the boundaries of the twofold absence and total concealment; hence the presence of those things occurrently present is said to reside in the juncture (*Holz* 327). Insofar as we turn towards presence itself, which here means the open realm within which all things arise, we also turn towards the juncture, which links presence and absence and thereby allows us a synoptic view of all that is unconcealed per se. To the extent that we turn towards what is occurrently present, the structural conjunction of presence and absence fades into the background; each thing becomes riveted in its own individuality, and the chaos of disjuncture prevails (*Holz* 327–28).

That which is occurrently present stands in the chaos of disjuncture; having arrived on the scene, it does not take leave immediately but persists for a while, insisting upon its own particularity. Yet the item can only persist for a *while*; only within this while can there reign the chaos of disjuncture. The while itself remains within the juncture. Since the while defines the span of presence which the item is allowed, and since presence is joined with the twofold absence, the occurrently-present entity must eventually acquiesce into remoteness. In other words each thing-present takes leave as its while or span of presence retreats and allows the total, conjoined structure of presence and absence to predominate.

Thus, although items persist, their persistence is more of a lingering; they come-to-be, they stay a while and they take leave. Because each thing-that-is must arrive on the scene and then take leave, it could be said that all items share a common fate. This common fate is no external career of a number of particulars; rather it links up each thing with the others in a much more fundamental way. By residing in the juncture, each thing-that-is acquiesces its own radical individuality and momentary occurrent-reality; for although the item resides within the disjuncture characteristic of the while, the latter remains within the juncture. An entity can only be in disjuncture to the extent that it is given an amount of latitude, a bounded realm of free play. At the same time this entity thus remains ultimately under the yoke of juncture.

Heidegger uses the term *Ruch* to characterize the way that the entities, in yielding their material, momentary presence, thereby stop opposing one another and instead adapt "one to another" in a structure of essences. *Ruch*, an obsolete German word, means something akin to *concern*. By this Heidegger does not mean to convey something merely mental or human; rather he is trying to convey a kind of relatedness fundamental to all categorical causality, logical combinitoriality and consciousness: a fundamental, causal principle of togetherness allowing for the mutual preservation of essences within the juncture rather than an annihilative opposition of actualities (*Holz* 331–32). To concern is to bear relevance, to allow affinity and togetherness.

The things-present, insofar as they are present, "let belong juncture and therewith also affinity of one to another." There are not two separate lettings here but rather one letting with a twofold effect.

Destiny and Ontological Difference 87

Something-present is present to the extent that it resides in presence, in the open realm of unconcealment. A thing can only be present by letting juncture belong, by acquiescing into juncture through the synopsis of coming onto the scene, tarrying, and taking leave—be it a rose, a flash of lightning, or a nation. To be present implies precisely this acquiescence into juncture and the three phases of coming onto the scene, tarrying, taking leave. Thus presence has the same scope as juncture. To be is to let presence belong; in residing in presence an entity comes to be alongside other entities, and this alongsideness allows the entities to bear relevance and affinity to one another. A flash of lightning in its eerie suddenness and a rose in its quiet beauty can come to be present when juncture is let prevail. In what sense do they bear affinity to one another? The lightning may illuminate the rose in bold relief. It may signal an oncoming rainshower which will refresh and nurture the rose, or a hailstorm, which will menace the rose. The rose and the lightning are both things which come to be and take leave from the same open realm. This affinity may seem trivial. Yet that is often the case with ontological matters measured by ontic standards; they only appear important in extreme situations. For example death puts the prince on the same footing with the pauper. This too appears to be an assertion of the poetic fantasy until it is time to die.

The things which come to be present do not let presence itself belong each to the other; the fragment merely says they "let-belong presence or juncture." The question arises, to whom do the things-present let-belong juncture if not to each other? Presence belongs, but it does not become the property of those-present. Heidegger tells us that juncture itself must belong to that in accordance with which presence, and therewith the overcoming of the chaos of disjuncture, prevails: *kata to chreon* (*Holz* 335). *To chreon* allows the occurrent-presence of those-present by compelling juncture; for only through juncture can those not now present ever arrive on the scene and become occurrently-present. Those present let belong juncture according to (the injunction of) *chreon*. Hence *to chreon* is Anaximander's name for the fundamental principle whereby all things are present; *to chreon* is the name for the Being of what-is, the name for the "relation" between Being and the things-that-are. *To chreon* is the incipient, original name for the Difference *as* Difference, that which grants not only things-present but presence itself (*Holz* 336).

Heidegger suggests that *chreon* is related to the verb *chrao*: to lend, supply. Thus *chreon* means a lending to things their presence such that the things thereby first arise as things (*Holz* 337). Heidegger translates *to chreon* through the German word *Brauch*; he resurrects the original meaning of this word which is our word *brook* in the old sense of *to use/enjoy*. At first glance this is a startling translation; we are in the habit of reading *to chreon* as *necessity*. *Brauch* in its contemporary usage means *custom, usage*—quite the opposite of *necessity*. Yet we must remember that Anaximander probably predates the separation of *physis* and *nomos*; indeed, it was the Sophists who first explicitly formulated this severance as a philosophical problem. Before that the ways of men, insofar as they were true and authentic, were considered rooted in something above and beyond man but yet not so automatic and mechanical as our word *necessity* might suggest. The kind of coming-to-be under discussion here is neither subjective and due to human custom and contrivance alone, nor mechanically objective where man would in no way be implicated. Rather the situation is something in between the mechanical automatic (*necessity*) and the humanly convenient (*custom*). I will coin the term *brookage* to express this *letting-be* which Heidegger has in mind in translating *to chreon* as *Brauch*.

Brookage, in lending things their presence, allots each thing its while, that is, its momentary duration in the occurrent present. In doing so brookage permits items to persist; insofar as they persist they have the tendency to insist upon their own presence and are thus in a sense, although bounded by the while, nevertheless momentarily in chaotic disjuncture. Since, however, their own presence is never delivered over them, juncture in the end prevails. Thus the chaos of disjuncture is never absolute; it has more the nature of a gap or hole within the juncture itself, a gap bounded on all sides however so that nothing can fall through it with finality.

Brookage, *to chreon*, is an incipient word expressing the Difference as a fundamental, dynamic principle which prevails not only over *those-present but over presence itself*. Something can only be present if it resides in presence, in the open realm; the total situation describing something present should be expressed: something-present (in its)-presence. This is, according to Heidegger, what is lurking in the participle *eon*, being (*Holz* 317–18; *WhD* 133–34, 148). What we most truly encounter is thus never merely the thing and never merely its presence but rather the twofold unity of both: the Difference as it unfolds itself in ensconcement.

This means that Being (meaning here presence) and the things-that-are cannot be separated from one another; they cannot be terms of a relation which stand over against one another. They are moments of the fundamental twofold principle as it unfolds itself in its twoness, where it allows the presence-of-things-present; that is, it allows things-to-be-present-in-presence.

Insofar as juncture prevails ultimately over the momentary chaos of disjuncture, brookage acts as a kind of gathering principle; brookage gathers back that which it has released for a while into the occurrent present by compelling those-released to let juncture belong. Thus those-released are gathered up into a preserved togetherness and affinity (*Holz* 340). Brookage is both *moira* (*fate* or *allotment*) and *logos*; as *moira* it allots to each his while, and as *logos* it gathers back all things into the juncture (*Holz* 340).

The adage of Anaximander, in Heidegger's translation, reads: "... in accordance with brookage; for they let juncture belong and therewith also bearing-affinity one to another (in overcoming) the chaos of disjuncture" (*Holz* 342). The "they" refers to those present, presumably both things and men. Then does this adage adequately allow for the unique ontological status of man, which is necessary to the Heideggarian framework? Or does the "they" lump man together with the things or impart a type of animism or "soul" to the things? This difficult question must be confronted if we are to see the full implications of the principle of brookage.

The first thing to remember is that the unique status of man is not defined by considering man versus things but rather by considering man as his essence is laid claim to by the primordial injunction, the Difference. Thus the fact that Anaximander did not explicitly speak of both things and men does not mean that the adage ignores the unique ontological status of man. The adage merely remains silent on this point.

We have already seen that man is by nature ecstatic, that he is in essence congruous with the open realm; this would also seem to indicate that his essence is congruous with juncture itself. Thus the letting-belong of juncture might occur at once with the essence of man: the dimension of human existence wherein each mortal is allotted his while, and therewith also placed alongside and in the midst of entities, where a mutual relevance and affinity prevails.

According to Heidegger's portrayal of man, the latter should have the greatest possibility of letting-belong-juncture and of falling into the chaotic disjuncture which reigns over the occurrent present. Man can become inauthentic and insist on his absolute presence; he can turn his back to the injunction of brookage and begin ordering the other things and other men about as he pleases. A tree is not capable of such rebellion.

On the other hand, even in our most fallen state, we are susceptible to the call of conscience, which reminds us that we are not the masters of the things-that-are. Our essence was granted also through brookage, and it is laid claim to by brookage (USp 155). Indeed, it would seem that we are most truly what is brooked, that is, what most truly arises through brookage: the letting belong of juncture and therewith also bearing-concern one to another.

Brookage cannot itself let juncture prevail; rather the fragment says: "according to brookage; they let juncture belong and therewith also bearing-affinity one to another..." "They" refers to those present: men, horses, trees, stones, the sea. Those present let belong juncture and therewith also bearing-concern in accordance with brookage. Two questions can be raised at this point: (1) When "they" let juncture ... belong, do "they" do so as a whole, or does each thing-present act as an individual? (2) What kind of compulsion is implied by the phrase "in accordance with"?

(1) In the discussion on "letting," it was seen that letting actually takes the emphasis away from the lettor because in letting the individual must acquiesce into the process. Thus in hammering, for example, a dynamic context prevails over the aggregate of individuals: hammer, board, nails, carpenter, to name a few possibilities. No one thing is responsible alone for hammering; rather "they" as a unified context let hammering occur. I would suggest a similar kind of interpretation in the case of letting juncture ... belong.

(2) The kind of "using" appropriate to brookage is one which first allows something to be; thus brookage grants, lends, supplies presence and therewith things-present (Holz 337). At the same time, since something-present must reside in presence, brookage limits things, bounds them within a while, a span of duration (Holz 339). Only through limiting can these things be preserved from falling into the absolute chaos of disjuncture. Thus to speak of brookage in terms of a relentless compulsion would be to present a onesided picture; for that which compels is also that which gives and embraces.

I have suggested that "they" let juncture ... belong in the same general way that "they" let hammering occur. The former occurs in accordance with brookage; it is not "self-motivated." In the case of the hammering we are accustomed to saying that man "motivates" the hammering. Man is the one who responds in accordance with a need for hammering; man is the "connecting link" between the need, the "in accordance with" and the letting-occur-hammering. So too I think it could be argued that man is the one who initally responds to the "in accordance with brookage"; somehow through man "they" let juncture belong ... and thus are granted their very presence. Man is, it would seem, the "special agent" of brookage, the essence or simple unity—signified by the "and therewith also"—of letting belong juncture and therewith also bearing-concern one to another. ... The "in accordance with" implies a kind of necessity not of a mechanical order but something to which man responds as he whose essence is congruous with juncture and therewith also bearing-concern.

The above interpretation is stated without documentation and proof, for its primary purpose is to give a few guideposts for what is to follow. Nevertheless the parallel should be obvious between this interpretation and man's relation to the Difference as the primordial injunction, whereby we are summoned to openly traverse the open realm and thereby encounter the things-that-are: whereby we are summoned to let-be-seen the to-show-itself. For in traversing the open realm we must go beyond the occurrent present in both directions; such a going-beyond necessarily involves letting juncture belong, since otherwise we would fall into a chasm. In letting juncture belong, the past and future-possible things are allowed an affinity with those occurrently present; all are assembled into a common, synoptic togetherness which is not biased by any one temporal mode. To state it another way, those things residing in the remoteness of absence are allowed proximity, and those residing in the proximity of the occurrent-present are allowed remoteness. Thus proximity and remoteness are distributed throughout the juncture of absence-presence.

Anaximander's fragment says nothing about the nature of man; it neither entitles us to an animistic view, where all things behave analogously to people, nor does it entitle us to the interpretation I have just given. Nor is there any hint of what kind of causation is involved in granting things their presence and in the "in accordance with." "Brookage" names the ontological difference in its elegant simplicity; but at the same time this simplicity remains enshrouded in ambiguity.

Logos: the ontological difference as letting-lie-before-and-gathering-back-up-in-preservation

We tend to think of *logos* as Reason or as a kind of world-ground or ordering principle. Yet if *logos* is to be considered as something so fundamental as to let all things be we must again be careful not to be misled into either extreme of mechanical or spiritual causality. Consequently Heidegger considers the word *logos* anew.

Logos comes from Greek *legein*, meaning *to speak, to lay* (German *legen* and *to gather* (German *lesen*) (VA 208). In laying something, that something is brought to a standstill, it is given a fixed position as is a stone or a brick. To lay is to-let-lie in the sense of bringing something to a place and letting it then lie there (VA 209). Heidegger suggests that this is the most fundamental meaning of *legein* and that all speech as *legein* is in essence a letting-lie. In speaking we bring something to lie before those who are listening; speech is a letting-lie-before (VA 212).

In bringing things to lie-before we let them lie together; we relate and combine. Letting-lie-together is a gathering, a harvesting. Harvesting implies more than mere accumulation, however; harvesting preserves, it ensconces (VA 209-10). *Legein* as *to lay* involves these three all at once: letting-lie, letting-lie-together, letting-lie-together-in-preservation.

Heidegger points out that harvesting does not imply merely a gathering-together of what is to be harvested; it implies also that harvesters gather together in the fields to carry out the work (VA 210). Harvest time is the time of festivity, of togetherness of men and bountifulness of food and drink.

An example of this threefold letting-lie is found in the notion of folklore. A number of deeds, details, and so on, are allowed to lie together in a song or poem. In this way an episode is preserved. But in folklore the preservation is not one of writing the episode down and hiding the manuscript. Rather the preservation consists exactly in letting the episode lie before, ever anew. In this way the song or poem is handed down through the ages; it is preserved.

From this example can also be seen why Heidegger goes on to say that involved in letting-lie-together, in harvesting and preserving is also a congregation, a gathering-together of people. The song or poem is recounted to others who can in turn pass it on. The assembly of people is essential to the preserving, for the latter can only occur if what is let-lie-before is gathered up. Yet mere gathering up in the sense of accumulation does not lead to preservation; implicit in the notion of preservation is that what is preserved can again be let-lie-before, that is, made accessible. When mere accumulation takes place the folksong or poem dies out; it is not passed on, it is not preserved.

Legein lets-lie-before that which is present; in other words *legein* does not itself bring forth from utter concealment that which it lets lie before. To see what is meant here we need only consider the fact that not all that is present lies before us. Speaking somewhat unrigorously, what lies before us is all that we can see. Seeing is not here confined to the physical sense however; what I see, what lies before me, is that with which I am concerned, that to which I am paying regard. There can be things present which I do not regard; but I am not likely to pay regard to something which is not before me. Thus something hidden can escape my notice.

Legein lets what is present lie before us. Thus *legein* controls and orients the horizon of what we take into regard. When something is present but does not lie before us, *legein* as speech can serve to bring this something before us. Heidegger uses in *Sein und Zeit* the example of the statement: "The picture hangs crooked on the wall" (*SZ* 217). Through speech someone can *let* the situation of a picture hanging crooked on the wall *lie before us*, although we had not hitherto noticed it.

What is let-lie-before is unconcealed; *legein* in its full sense of letting-lie-before in togetherness and preservation is a bringing into unconcealment and preserving this unconcealedness. Thus *legein* is conditioned by and subservient to unconcealment; *legein* is a kind of instrument or process whereby, in letting-lie-before the concealed is brought out into the open. The general process of divulgence, whereby the concealed is brought into unconcealment, constitutes the dynamic character of presence. Since presence has been allied with the Being of the things-that-are, it can be said that speech as *legein* derives its essential character from the Being of the things-that-are rather than from any such thing as meaning, reference or articulate sounds (VA 211–12).

After defining speech as a letting-lie-together in preservation the question arises as to how we should regard hearing. True hearing has already been allied with a heeding rather than with the perception of sound. Heidegger interprets hearing as *homologein*: to let that which is allowed to lie before us *lie intact in the self-sameness* of its original lying-before (VA 215). The original lying-before is, in everyday terms, the context in which something was disclosed as it was brought before us.

For sake of clarity we can contrast true hearing, that is, *homologein*, from half-hearing, whereby we merely let what is allowed to lie before us *remind us* of something *similar* which we then proceed to formulate, that is, to let lie before us, instead of that which was originally let lie before us. Instead of letting the original self-same disclosure lie before us, we settle for something similar. True hearing implies that we let what is allowed to lie before us *remain* in its original lair, that is, the context of its presence; we recognize that it was not our own speech which allowed the item to lie before us; we heed what is disclosed *as* it is disclosed.

Legein refers to the speech of mortals. Mortal speech can only let-lie-before something which is already present in some way. Mortal speech can bring into proximity that which is remote, but it cannot bring something out of radical concealment into unconcealment. It can only let-lie what is delivered over to it, what is already present in some way. What is present resides in presence; Heracleitus interprets this original lair in which things lie as the *logos*: the pure, incipient to-let-lie-before-together-in-preservation. *Logos* is the fundamental *to lay* whereby what is present lies of its own accord before us. True hearing lets lie in its original *logos* that which of its own accord lies before us (VA 216–17).

Thus true hearing as *homologein* remains a *legein* which can only let-lie what already lies-before in its original, incipient lair, that is, its original context of disclosure; what already lies-before never originates in *homologein*, nor in *legein*, but rather in *logos*. Although *legein* and *homologein* are delivered over to man, the original, incipient to-let-lie is not in his control; man does not himself bring forth the objects from a radical concealment, but must always approach them in a context of their disclosure, that is, in their original lair (VA 216–17).

In the case of true hearing, we let something lie intact in its original lair, we do not set ourselves off against it. We let ourselves belong in harmony with *logos*; we let our *legein* correspond and conform with *logos* (VA 217). When this happens, Heracleitus says that *sophon* occurs. Heidegger translates the word *sophon* as *das Geschickliche*, a word related to *Geschick* (fortune, fate, skill), to *Geschichte* (history, but having the original meaning of something like event, case, circumstance) and to the verb *geschehen* (*to happen, occur*). Unfortunately there is no single English word which covers all of these meanings. However the word *history* comes from a Greek word meaning *learning by inquiry, knowledge, narrative*. The word *histor* meant *learned, wise man*. The word *destiny* also covers a number of the meanings enumerated above; *destiny* also implies adeptness, insofar as when we say someone is destined for a certain thing we already assume that he is adept. Thus I propose as a translation of *das Geschickliche* the compound term *destinate-historical*.[5]

Destiny and Ontological Difference   95

Heidegger never explains in much detail what is meant by the destinate-historical. We can assume that at least part of what he means is that a kind of synoptic manifestation occurs, such that all events and things are seen as belonging to a common presence embracing not only what is coming-to-be and what-is but also what-has-been. In the words of Anaximander, the letting-belong of juncture occurs, whereby all things are seen as belonging to a common presence. The granting of this common presence is the occurrence of the destinate-historical. Letting juncture belong is a letting-lie in the original lair of the common presence; thereby the destinate-historical occurs, whereby all things bear a mutual relevance and importance. This parallel with Anaximander's adage shows us that in the doctrine of *logos* Heracleitus is concerned with how letting-juncture-belong takes place. He looks at it as a conforming of our mortal *legein* to the original, incipient *logos*.

When we conform to *logos* the destinate-historical occurs. Heracleitus tells us that the latter manifests itself as *hen panta*: One-all (VA 219). Only when we let ourselves belong to *logos* do we experience the destinate-historical; and only then do we experience the *hen panta* (VA 221). The *hen panta* allows the synoptic view whereby all things coming-to-be, occurrently-present and having-been are assembled into one, common presence. Such a synoptic view is necessary for the letting-be-seen; hence we let-be-seen when we let things lie intact in their original lair, when we conform to *logos*. This we do through authentic encounter, where we show active concern for things; we bring them forth as in the case of poetry or artifacts, while at the same time not asserting our own will and mastery over these things. We let them lie in the lair of presence rather than in our own domain.

Hen is the unifying, gathering fold. It gathers *panta*, all the things-that-are, up into one presence, one fold. Yet *hen* does not merely relate and bind together so that the result would be an amorphous mass; rather different things are let lie together, for example, contraries such as day and night, hot and cold. But how can two opposing things be let lie together?

Logos as the pure, incipient to-let-lie lets all that is lie in the original lair of presence; *logos* thus allows things to lie in unconcealment. Now day and night oppose one another such that when one is present the other must be absent. In letting day and night lie in unconcealment, *logos* must bring these two apart and let them lie separately. Yet this "separately" bears an important qualification; for the contraries must be separated specifically by the poles of presence and absence, so that when one is arriving the other is departing and vice versa. Although day and night must occur alternately, they are let lie together in a common presence which governs their alternation in occurrent-presence and absence. If day and night did not lie together in some common presence there would be no juncture; we would not recognize them as alternating contraries, where one is absent just when the other prevails in occurrent-presence. There would be no continuity from day to day; there would be nothing like memory or even thinking, for we would be confined totally to the immediately present.

Hen panta gathers up all that is let-lie so that a continuity or juncture prevails rather than the chaos of Anaximander's disjuncture. *Hen panta* also brings-apart, that is, things are released into the occurrent present and given freedom within a time span.

If the mortal *legein* adapts itself as a *homologein* to the *logos*, that which is present is allowed to lie intact in its original lair. The destinate-historical occurs; a continuity of presence prevails so that all things lie in a common relatedness. If in speaking we do not let things belong to the original *logos* this continuity dissolves, and what-is appears as a number of isolated entities complete in themselves. Once this happens, the *All* or *panta* is viewed as a plenum ordered externally to the One (*VA* 224).

Logos is the original, incipient letting-lie whereby things arise and hence reside within the lair of presence. *Logos* prevails as the destinate-historical, the *hen panta*. *Legein* is a finite letting-lie, whereby things-present are let-lie-before so that they may be seen and preserved in their unconcealment. In letting-lie-before, the finite *legein* must always adapt itself to the original lair of presence, the context of disclosure; for only those things in the original lair can be let-lie-before. The *legein* must adapt itself to *logos* in order to abide as *legein*. To the extent that this adaptation, this *homologein*, takes place, the destinate-historical occurs. Hence the adaptation is also necessary for *logos* to prevail. *Logos* and *legein* exhibit a mutual dependency.

At this point a difficult question must be raised. *Legein* is mortal speech, and we cannot deny that to have mortal speech there must be mortals. Mortals have apparently been let lie in the original lair of presence along with other things. When he dies that particular, finite *legein* also ceases to be. Was then the mortal *legein* also let-lie-before? But if it is needed (namely in its adaptation to *logos* through *homologein*) in order that *logos* prevail, how could the latter have let it lie before? Or, if *logos* can first let *legein* lie-before, why should the former be in any way dependent upon the latter?

The answer to such a question must lie in the apparent fact that *logos* is not exhausted by the mere release of things into their unconcealment (presence)—at least not exhausted if we interpret this release as something mechanical. In letting-lie is implied a bringing to a fixed position, a bringing to rest, a kind of stilling action. In letting-lie there must be a place for the object to be let-lie; thus the incipient to-let-lie must provide a lair for objects to lie in as well as release the objects from concealment. This lair is not a mere receptacle however; it is not independent of the things that are within it. The receptacle is in a sense the togetherness of things; but this togetherness is not a post hoc but rather a condition that things be at all. Moreover the togetherness of things is their unconcealment; for in the divulgence of concealment is let-lie at once the lair of presence (unconcealment, togetherness) and the things arising therein.

Although *logos* provides the lair of presence and therewith the things-which-are-present, something must be *seen* before it can be truly present. This is the part of the mortal *legein*: letting-lie-before-(so that all may see)-intact-within-the-original-lair. The mortal *legein* deals with and acknowledges the presence of the things; it lets them be preserved and lets them belong to the original lair, rather than letting them fall into chaos because of their releasement. This the mortal *legein* can do because it is present alongside these objects; this the *logos* cannot do as such if it is to embrace and bound all things rather than set itself over against them. For acknowledgement implies a vis-à-vis, not a mere enveloping, of the thing to be acknowledged.

The release of things into unconcealment thus in a way estranges these things from *logos* as such. In order to retain a hold over the things *logos* must become other to itself and yet still retain its own radical self-sameness as *logos*. This otherness is accomplished through the mortal *legein*, which is both other than *logos*, since it is mortal and also alongside the things. and nevertheless congruous with *logos*, since it is letting-lie-before-together-in-preservation. *Logos*, in order to keep a hold on the released things, must deliver itself over to *legein*; this deliverance from its side prevails as the *hen panta*, and from the side of *legein*, which accepts the deliverance and adapts and conforms as *homologein*, the deliverance prevails as the destinate-historical, that is, the authentic, synoptic vision.

Through the mortal *legein* things are let-lie-before, gathered back up and passed on. Often they are passed on however without regard to the original lair of presence, without regard to the original significance and context of disclosure. Thus the lair is forgotten and only specific items are passed down. The preservation through tradition is a dissimulation where the incipient disclosure of what-is in its lair of presence is replaced by shallow repetition. The things, instead of being gathered back up and let belong to the original lair, are further dislodged. Yet the fact that the *legein* is mortal insures the eventual gathering up of all things back into the original fold, even though this be at the expense of their presence and at the expense of the things themselves. Death requires the relinquishing of man's claim on the things-that-are. Without the mortal *legein* the things delivered to it cannot be present unless they have been let lie before others, who can then continue to preserve but also to dissimulate. Those things which are not let lie before others are relinquished and gathered back into the fold of concealment. Mortality is, as it were, the final trump card which is held over the things which have been released into unconcealment and over man's inauthenticity and rebellion in refusing to let these things belong to the original lair of presence.

To sum up, *logos* is a statement of the fundamental twofold principle whereby things-in-their-presence are allowed to arise and lie-before. The *logos* is an expression, a destinate-historical utterance, of the Being-of-the-things-that-are (VA 227–28). Yet this original sense of *logos* itself did not endure. Not only did the things-which-are-let-appear get dislodged, but the lair itself drifted and shifted. The lair itself cannot be taken for granted but needs continual preservation, much as a house needs upkeep and repairs. But preservation and upkeep are also what bring about dissimulation; what is falling away into decay and oblivion can never be brought back as the self-same which was. This is the meaning of the drift, the bringing-apart of the opening within the lair itself. Yet somehow the primordial call is able to span or gather up, to be sure in ensconced fashion, what gets set adrift. But this is the theme of *moira*.

Moira: the ontological difference as apportionment

Parmenides mentions *moira* in a fragment which deals with the self-sameness (*to auto*) of Being and thinking; Heidegger quotes the fragment in Kranz' translation which might be rendered in English: "Thinking is the same as the thought that Is is; for you cannot find thinking without the things-that-are, in which thinking becomes explicit (in utterance). There is nothing nor will there be anything else outside that-which-is, since *moira* has bound it so as to be whole and immovable" (VA 231).[6]

Heidegger calls our attention to the fact that the term "things-that-are" means the things in their presence and thus has a dual meaning conveyed by the Greek word *eon* (VA 240). This latter word is a participle rendered in both meaning and form by our word *being*. All participles have for Heidegger a dual character. He uses the example *blooming*, which can either mean *that which blooms* (*ein Blühendes*) or the *dynamic nowness* of being in bloom (*im Blühen, blühend*) (WhD 133). In the same way the word *being* (*Seiendes*) can either mean *what-is-being*, that is, a being, or it can mean the *ontological dynamic* whereby those things-which-are can be: they are qua being. Yet this latter meaning of *being*, which I will hereafter signify by the term *be-ing*, is not to be thought of as any kind of "act" in which an entity engages, nor is it merely the "that-ness" of a "what." Only through be-ing can a being first arise as something (a "what") which is. The term *thatness* or *existentia* is traditionally applied to an entity already present in its whatness; "that it is or is-not" is only relevant to an entity which has already arisen. The attempt to use the distinction between whatness and thatness, or essence and existence, as one's point of departure confines itself to the entities and suppresses the problem of be-ing: the problem of that by virtue of which things first arise (2 Ni 458).

Thus Heidegger interprets Parmenides' fragment as saying that thinking is not to be found except within the "twofold," that is, within the domain of being(s)/be-ing (VA 242). There is an ambiguity in this statement which is easily overlooked. The statement tells us that thinking belongs within the domain of the twofold, but it *does not say why*. Specifically it does not say that thinking belongs within the domain of the twofold *because* thinking is something-present. Indeed, were this the main reason, Parmenides could have said so and would have no cause to deal explicitly with the inner relation between thinking and the twofold (VA 242, 249–50).

Thinking occurs within the domain of the twofold. This means that thinking occurs on account of and by virtue of the twofold, for the latter compels and allows all that is within its domain (VA 242). Heidegger appeals to Fragment VI to discover something more of the necessity that thinking belongs within the domain of the twofold. This fragment, in Heidegger's translation, reads as follows: "It brooks [that is, uses necessarily and thereby allows to arise in its full essence] letting-lie-before (*legein*) as well as also taking-into-regard [*noein*: thinking in the narrower sense, apprehension]: being(s) being" (WhD 136).[7] The construction *it brooks* is an attempt to render a medial form, a kind of action which does not owe its presence to a definite agent, but has a priority such that it would first give rise to any kind of "what" which we might take to be a phenomenal agent. The "it" then refers to the twofold, but not merely as an agent engaged in an external action; rather the twofold *abides* as the "it brooks": an abiding, dynamic principle which calls forth letting-lie-before and taking-into-regard (WhD 115–17, 136).

Thinking, when summoned forth, comes to be present; but the presence in this case is not the presence of a "what." Rather the presence itself is something highly dynamic in character. Thus Heidegger tells us that the presence of thinking is "under way" to the twofold of presence and things-present (VA 242–43). Thinking is neither pure presence nor is it something-present in the sense that a thing is something-present. Yet what alternative status could thinking have? It cannot be the same as the twofold, for the former is summoned forth by the latter.

The term "under way" does not mean to suggest any kind of "becoming" if we mean by the latter a kind of linear passage of "nows" such that these "nows" are only ordered but not bounded. "Under way" refers to the traversal of the open realm; it need not be linear, since in traversing we can distinguish things within the open realm and then assemble all these things into one synoptic apprehension. The traversal "under way" is always bounded, for it was seen earlier that the traversal of the open realm involved an ecstatic going-beyond and then a return. Thinking has the structure of a dynamic, reflexive relation rather than a linear type of becoming.

In his discussion of Parmenides Heidegger tends to assign the twofold two modes or moments: (1) the twofold as a simple, dynamic domain or "fold," that is, brookage, and (2) the twofold in its "unfolded" moment as presence/things-present (WhD 148). Regarding the fact that thinking is "under way" Heidegger tells us: "Thinking abides in presence due to the twofold which remains unsaid. The presence of thinking is underway to the twofold of Being and what-is" (VA 242).[8] To be noted is the simple use of "twofold" (*Zwiefalt*) in the first sentence in contrast to the phrase "twofold of Being and what-is" (*Zwiefalt von Sein und Seiendem*) in the second sentence.

Heidegger uses simply "twofold" when he speaks of that on whose account thinking occurs.

Elsewhere Heidegger makes the following statement concerning the presence of things-present: ". . . their twofoldedness out of the concealment of their simple fold harbours the injunction" (WhD 148).[9] "Their" twofoldedness, that is, the twofoldedness of presence and things-present, involves a concealing of their simple fold. In the unfolding of the simple fold towards the twofoldedness of the twofold (which is at the same time a concealing of the simple fold per se) is contained the injunction that thinking be summoned forth. Thinking originates in the unfolding (and concealing) of the simple fold toward the twofoldedness of presence/things-present. Hence the very occurrence of thinking is "under way to the twofold of Being and what-is," that is, under way to the twofold in its unfolded moment.

This latter statement has two possible meanings: (1) thinking enjoys a movement parallel to and isomorphic with the unfolding; hence thinking has primarily the function of disjoining or severing the twofold into its twoness of presence/things-present. The severence shows up in the fact that the things become openly manifest, whereas their pure presence, that is, the open realm which allows them a place to reside and be encountered, tends to conceal itself. (2) Thinking has the function of "healing" the fissure created by the unfolding, thus allowing the open realm of presence and the things-present therein a return from their twoness back to the twofold. This would be done by assembling the things-present into the common, synoptic presence; for in this assemblage the belonging-together of presence and things-present is stressed rather than either of these two poles. The return back to the twofold is never literally brought to completion, however, since thinking is always "under way."

Heidegger interprets Parmenides as saying further that thinking becomes explicit or occurs through utterance (VA 231). This suggests a connection between language and the unfolding of the twofold; this connection has already been prepared for in the discussion on *logos*. *Logos*, it will be recalled, is the pure, incipient letting-lie whereby things are brought to rest in the lair of presence. To the extent that things are allowed to arise, *logos* has the character of a divulging out of concealment; to the extent that these things are brought to lie in a synoptic presence *logos* has the character of an assemblage or gathering-back. Once the things are allowed to arise they are released from the incipient fold of concealment, that is, they are divulged. Now *logos* qua *logos* cannot be together with the things that are divulged, for it cannot divulge itself and still retain a radical self-sameness with itself. To put it more simply, *logos* cannot divulge itself and still remain *logos*.

As such *logos* cannot qua *logos* assemble the things-present (that is, the things-divulged), for what assembles must in some way be together with what is to be assembled. This is exemplified, for example, when a leader of people must invoke a feeling of togetherness between himself and those he leads; only by doing so can he keep their support, that is, keep them assembled as a group under his leadership. *Logos* therefore calls the mortal *legein* into presence; the latter then resides in the midst of things and can assemble them. *Logos* then gathers the mortal *legeins* into the simple fold of concealment. Let us now see what light is shed on this situation by the fragments of Parmenides which are under present consideration.

Heidegger interprets the meaning of *utterance* (*pephatismenon*) as founded in *phasis*: to bring into the foreground, to allow to appear (VA 244). "*Phasma* is the appearance of the stars, the moon, their coming forth to shine, their self-concealing. *Phasis* names the phases. The alternating ways of its shining are the moonphases. *Phasis* is saga; to-say means: bring forth to shine" (VA 244).[10]

Noein, or taking-into-regard, occurs through utterance, that is, through being brought forth. However *noein* is not brought forth in the way that a thing is brought forth and let lie in presence; rather, *noein* is brought forth under way toward a synoptic apprehension of things-present in their presence (VA 245). On the other hand *phasis* as letting-appear seems to parallel the unfolding of the simple fold into the twofold of presence/things-present. Thus Heidegger makes the statement: "The unfolding of the twofold presides as *Phasis*, to-say as bringing forth to shine" (VA 253).[11]

Phasis does not itself divulge things from their concealment but rather primarily brings the *presence* of things into illumination. *Phasis* grants the illuminative clearing (the open realm of presence) wherein things can appear. It is above all the coming-forth of this illuminative clearing (*lichtendes Scheinen*), in whose light the things can arise and appear, which most truly marks the unfolding of the simple fold into the twoness of the twofold. This unfolding was experienced (but not necessarily recognized as such) by the Greeks as divulgence (*Entbergung, aletheia*) (VA 247).

Phasis, in unfolding the twofold, calls forth the mortal *legein* (letting-lie-before) and *noein* (taking-into-regard), ordering these

Destiny and Ontological Difference   103

two in an interdependence such that the former lets things lie before the latter. Noein takes these things into regard, assembling them and preserving them, while under way to the twofold of presence/things-present (VA 243, 245).

The occurrence of the illuminative clearing and the coming-forth of thinking are not two separate events however; rather they are said to be the "self-same" (to auto). This cannot merely mean that these two are identical, for sheer identity would not adequately imply that thinking is *needed* in the unfolding. Heidegger calls to auto a Rätselwort (VA 241). In this word Parmenides remains silent about the "self-same" which unfolds the twofold and allows the assembling apprehension to arise and proceed "under way" (VA 249). The coming-forth of the assembling apprehension is then the "self-same" as the occurrence of the illuminative clearing.

On the other hand the inner structure of legein and noein (Anaximander's to chreon, Heraclitus' homologein) lets things lie-before in the light of the illuminative clearing; as such, these things are released to allow the possibility of an assembling apprehension. Thus noein can give the semblance of being "outside" the twofold, in the twoness of the unfoldedness and always merely under way to the twofold itself. Yet this semblance of being "outside" is exactly what must be qualified, for the inner structure of legein and noein remains under the yoke of the twofold: "However divulgence grants the illuminative clearing of presence, in so far as, if things-present are to appear, divulgence uses letting-lie-before and also taking-into-regard and by thus using thought holds it within the purview of the twofold. Therefore there can in no way ever be something-present outside of the twofold" (VA 251).[12]

It is not until all this background is laid down that Heidegger finally addresses himself to the problem of moira. It is, according to Parmenides, because moira has bound the twofold so as to be whole and immovable that nothing can be present outside the twofold (VA 251).

Moira is the apportionment (Zuteilung) which vouchsafingly allots and thereby unfolds the twofold (VA 251). Our translation of moira is "fate," "destiny"; the unfolding of the twofold is the unfolding of the destinate-historical: "[Apportionment] . . . is the self-assembled and thus unfolding dispensation of presence as presence of things-present. Moira is the fated bestowal of 'Being' in the sense of eon" (VA 252).[13] Eon, being, refers to the simple fold of the twofold rather than its explicit twoness of the unfolded mode. Hence Heidegger tells us that *moira* has released *eon* into the twofold, that is, into the unfolded mode (VA 252).

The releasing of *eon* into the unfolded mode of the twofold is at the same time a binding "so as to be whole and immovable," since the unfolding is only into the twoness of the twofold and not an endless, linear process. In this releasing and binding occurs the presence of things-present (VA 252-53).

It will be recalled that the unfolding of the twofold prevails as *phasis*, "to say" in the sense of bringing-forth-into-light (*zum Vorschein bringen*). In the unfolding, thinking as the conjunction of *legein* and *noein* is summoned forth to preserve the things-in-their-presence. But this is tantamount to preserving the twofold itself by letting the things belong to presence and letting the latter belong to that which originally granted it. Insofar as this is done "thoughtful saying" occurs, where we heed the presence of things-present rather than engaging in idle talk.

In the unfolding of the twofold the illuminative clearing shines forth, in whose light the things appear. The twofold itself, that is, the simple fold whereby the twofold remains always a twofold when unfolded rather than becoming a radically severed duality, remains concealed. As Heidegger puts it: "In the fateful bestowal of the twofold, however, only presence attains the status of shining forth and only things-present come to appear. The destinate bestowal retains in concealment the twofold as such and all of its unfolding" (VA 252).[14] Because the twofold itself remains concealed, the mortals do not recognize that their own thinking has been summoned forth. Thinking comes to be regarded as a natural faculty, something taken for granted. Men confine their concern to what immediately importunes without regard to presence itself. Mortal speech degenerates into idle talk, dominated by words and the mere saying of names, because men do not heed that which summoned forth their thinking and do not allow their speaking and taking-into-regard to conform to the original lair of presence (VA 254). Hence Parmenides' Fragment VIII continues: "There is nothing nor will there ever be anything else outside the things-which-are, since *moira* has chained it so as to be whole and immovable. Therefore all that the morals have established, although they are convinced of its truth, is governed by mere names: 'coming-to-be' as well as 'passing-away,' 'to-be' as well as 'not-to-be' and 'change of place' and 'alternation of brilliant colour'" (VA 231-32).[15]

Everyday thinking comes to be governed by words which already seem to exist and seem to name already existing things—things which are immediately taken to be true. All apprehension becomes hasty and fleeting, so that one is driven from one thing to the next, fleetingly perceiving one thing *as well as* another. Obvious to the lair of presence, the true place in which all things-present reside, man sees only a continual flux and change of place. Blind to that which summoned forth mortal thinking, the latter has no one thing to hold on to, nothing by which to orient itself. Everyday thinking in its restlessness is the victim of a phenomenal reality whose "truth" lies in "alternation of brilliant color" rather than in the shining-forth of the illuminative clearing, the quiet radiance of the unfolding twofold (VA 254–55).

In the unfolding of the twofold into presence/things-present the twofold qua simple fold, that is, that which grants the unfolding, remains concealed. The actual unfolding is at once the divulgence of presence/things-present and the concealing of the simple twofold: "Then does there reign implicit within divulgence its own concealing? A daring thought. Heraclitus actually thought it. Parmenides experienced this thought without thinking it" (VA 255).[16]

The unfolding of the twofold: aletheia

We have seen earlier that Heidegger does not view entities as discrete givens which have always been there or which, after having come-to-be, are no longer problematic. Rather entities must arise within the open realm and continue to reside therein in order to be the things they are. The very being of an entity is its being-allowed to continually arise or appear within the open realm. More important however is the fact that the open realm itself is not viewed as a static container whose presence is nonproblematic. Rather the open realm is first allowed to come forth and prevail as pure presence through the unfolding of the twofold.

In his earlier works Heidegger spoke of *ontic truth* and *ontological truth*; the first term referred to the unconcealedness of things qua things, whereas the second term referred to the unconcealedness of the open realm of pure presence (WG 12–13). Regarding these two layers of truth he says: "They belong essentially together on the basis of their affinative relation to the Difference between Being and things-that-are (ontological difference). The two-pronged essence of Truth which comes about in that manner is only possible at once with the eruption of this Difference" (WG 15).[17] This is the first time Heidegger mentions the ontological difference explicitly. He says that the ontic-ontological essence of truth is founded upon the ontological difference, insofar as the former is only possible "in eins mit dem Aufbrechen dieses Unterschiedes."

This means that if we merely confine our attention to the ontic-ontological distinction and inquire only into the relation between the ontic and the ontological, between things and the open realm, we are not striking at the heart of the ontological difference. What we must do is inquire into the *occurrence* of this two-pronged essence of truth: the *unfolding* of the twofold into presence/things-present or, in Heideggerian truth-terms, into unconcealedness/things-unconcealed.

The unfolding of the twofold is a divulgence or disensconcement (*Ent-bergung*); this Heidegger identifies with the incipient *aletheia*. *Aletheia* is the occurrence of presence/things-present. Since, however, things can only be things qua arising and residing in the open realm of presence, the fundamental character of the unfolding is the inception of presence itself. This appears to be the necessary and sufficient condition that things come to be. We read for example in *Was ist Metaphysik?*: "Regardless of the way in which things-which-are might be laid out for interpretation ... the things-which-are appear as things-which-are always in light of their Being. Everywhere, insofar as metaphysics represents to itself the things-which-are, Being has shone forth as illuminative clearing" (WM 7–8).[18] Elsewhere we read: "Being must in fact already at the outset shine forth of its own accord and on its own terms in order that the things-that-are can appear at any particular time" (SvGr 111).[19] The arrival of Being as the open realm of presence, the breakthrough of the illuminative clearing, marks the very crux of *aletheia*.

The difficult essay of Heidegger's entitled *Aletheia* is an attempt to deal with the unfolding itself, with the inception of the open realm of presence. Obviously the open realm cannot shine forth as does something which is, for example, the sun. To characterize the way in which the open realm can break forth and abide and yet not *be* as something-which-is, Heidegger uses as his point of departure the Heracleitian fragment: "How can one ensconce himself before that which never sinks away?" (VA 259).[20]

Heidegger first asks, what is implied in the phrase *ensconce oneself*? Shall we thereby understand an active concealing, as when we hide ourselves before an enemy or when a sinful Christian tries to hide before an omnipresent being? Does this fragment suggest that there is something which sees all, so that it is impossible to hide from it?

Heidegger maintains that the Greek term employed here (*lathoi*)

Destiny and Ontological Difference 107

is not an active form but rather one signifying a remaining-concealed (VA 262). The emphasis is upon a state or condition rather than upon a voluntary act. If we remain concealed vis-à-vis something, this something sinks away from us and we from it. This appears to happen, for example, when we forget something. Forgetting is not, strictly speaking, an act. We cannot pin down a moment of time when we forget something, just as we cannot pin down our last moment of being awake before we drop off to sleep. Because forgetting has no discrete beginning, in forgetting we have already forgotten, that is, we are already in a state of remaining-concealed.

Our human forgetting is permeated with finitude; we can be oblivious to that which is nevertheless present to others. When we stumble over a stone we do not assume that the stone suddenly loomed up out of radical concealment; the stone has the character of having been in a sense already there. The stone is familiar to us; indeed, were this not the case we could not recognize it as a stone in our sudden encounter of it. Our forgottenness of the stone is somehow intertwined with a familiarity with the stone. Evidently human forgetting does not suggest a radical concealment; human forgetting is confined to the open realm.

This fact is underscored by the fact that Heidegger views everyday man in a state of forgottenness and yet positively related to a prethematic disclosure of Being. This vague familiarity with Being, which structures and guides our preoccupation with the things-that-are, was called the *Seinsverständnis*. The *Seinsverständnis* permeates all levels of human existence; hence "no one can remain concealed to that which never sinks away" (SvGr 146; VA 266).

Heidegger addresses his attention next to the phrase *that which never sinks away, to me dynon pote*. According to him *dynon* means to sink down into something and thereby become concealed. To sink away is to sink into concealment: "Sinking down as conceived by the Greeks occurs as passing-into concealment" (VA 266).[21]

But what sort of thing never sinks into concealment? The entities arise, but they also retreat into remoteness and obscurity. The world of entities is a phenomenal one, infected with change and flux. Clearly that which never sinks into concealment must be "something" which steadily arises. This can only apply to pure emergence itself: *physis* (VA 267).

Pure emergence is not a mere developing or one-way movement from potency to act. Pure emergence and never-sinking-into concealment are one and the self-same (VA 270). Inherent in the very nature of *physis* is an essential, albeit obscure, relation to concealing and concealment. *Physis* is in some sense a revealing/concealing.

The inner belonging-together of revealing and concealing is expressed in the Heraclitian fragment: *Physis kryptesthai philei—Physis loves to hide* (VA 270). Heidegger interprets this as meaning that pure emergence is inclined toward self-ensconcement. The self-ensconcement is not a mere hiding however; rather it is also a preserving which allows the pure emergence to retain its character and not exhaust itself. On the other hand there can only be a self-ensconcement on the basis of a propensity for emerging (VA 271).

The mutual inclination of *physis* and *kryptesthai* does not at all have the character of a Newtonian attraction between two already given particles. Rather the mutual inclination spoken of here is so fundamental that it first allows *physis* and *kryptesthai* to abide fully. This mutual, permissive inclination Heidegger calls *Gunst*, favor: *philein* and *philia*. Accordingly he retranslates the fragment *Physis kryptesthai philei*: "The emergence (out of self-concealing) is allotted to self-concealing by favor" (VA 271).[22]

Far from being opposed to one another, revealing and concealing exhibit a fundamental, inner membership. Revealing does not conquer over concealing but actually needs the latter in order to abide as pure emergence (VA 272). Rather than having the character of a continuous process of development, *physis* is a reflexive, mutual interplay of revealing/concealing. As a never-sinking-always-into-concealment *physis* is a continual emergence out of concealment (VA 272).

Heraclitus uses the word fire to further describe this pure emergence out of concealment. Fire, prior to both gods and man, is the world itself, the dynamic, open realm: "World is enduring fire, enduring emergence according to the full sense and import of *physis*" (VA 275).[23] The Heraclitian fire is not merely a wild flaming or burning, although it is forbidding. Fire is described as ". . . the illuminating governance, the guidance, which gives boundaries and measures and also withdraws them" (VA 275).[24] Although Heidegger does not explicitly say what is meant by *Masse* here, I think we can safely assume that measures and boundaries are meant which would make possible articulate intelligibility. The limits of intelligibility are precisely the limits of the proximate domain of fire. A fire during a black night gives measures and boundaries so that we can distinguish objects within the fire's proximate domain. The fire opens up a space,

Destiny and Ontological Difference 109

for it consumes everything which does not respect its boundaries. The fire flickers, and in flickering resides the possibility of withdrawing these measures and boundaries. Heracleitus called fire *das Sinnende* according to Heidegger, that which muses. The Heracleitian fire assembles all and ensconces all. Thus Heidegger identifies fire with *logos* and also with *Lichtung, illuminative clearing* (VA 276).

Fire as the shining-forth of the illuminative clearing does not merely brighten; it releases things from the darkness of concealment, allowing them to arise and be assembled in presence. Thus Heidegger says that to-illuminatively-clear (*Lichten*) is "musing-assembling bringing-forth into the free and open, is granting of presence" (VA 276).[25] The shining-forth of the illuminative clearing is disensconcement par excellence: *aletheia*.

It is important to remember however that the disensconcement resides in an ensconcement, albeit one which finds its own essence in disensconcement (VA 276). These two belong in an inner unity such that each one can only abide through the allowance and permission of the other. This is why, for example, the ancients saw dynamic principles as contraries, that is, as a continual alternating of advancing and retreating of two elements, each delivered over to the other within the domain of the revealing/concealing. The advance of day, for example, is the retreating of night. Yet day does not radically change into night; were this the case, day could not appear again, that is, there would be no continual alternation of day-night-day. . . . Viewed phenomenally, day does sink away, but only into a remoteness and not into a radical absence.

Ensconcement or concealment is not some kind of underlying, hidden, indeterminate, inert realm such as a void, a prime potency, or inert matter. Since ensconcement can only prevail in connection with disensconcement, the former must be as dynamic as the latter. These are not just another set of contraries, for revealing and concealing do not alternate vis-à-vis one another. These two belong together in an inner self-sameness; they are one and the same occurrence (VA 270).[26]

The inner self-sameness of revealing and concealing is called the "realm of all realms," wherein the simple unity of the One (*hen*) resides (VA 272). Here all things arise, waxing together (*concrescere*). The realm of all realms is pure concreteness (VA 273). We can ally this realm with pure proximity and with the simple fold ensconced within presence/things-present (WhD 134–36).

The open realm of presence is not a mere static horizon of already-revealedness which faces us, much as a "transcendental realm" might face a subject (Gel 38–39). Indeed, the revelatory horizonal character of the open realm is not something unambiguous and unequivocal; this revelatory aspect of openness is only one side of the revealing/concealing. The other side is reflected in the forbidding emptiness and the mysterious bounded boundlessness of a realm which seems to have no bounds and yet for this very reason functions as the most forbidding boundary of all-pure nothingness.

Heidegger next calls our attention to the fact that the Heracleitian fragment asks how *someone*—not something—could remain concealed to that which never sinks away. The "someone" need not be merely a man, however; "someone" could also be a god. The steadily emerging illuminative clearing lets both gods and men reside in the open realm of presence such that never might someone of either of these remain concealed. Nevertheless each of these two resides differently in the open realm. Whereas the gods are *die Hereinblickenden*, those who see into the illuminative clearing itself, men must reside in the midst of the things, and these are accordingly more accessible than the illuminative clearing (VA 277).

Yet what about the other kinds of things-present, such as trees, stones, the sea, animals? Does the Heracleitian adage, in using the word "someone" have only a limited application since it speaks only of men and/or gods? Heidegger suggests that, far from limiting, the adage might actually widen the scope of applicability by singling out in a special way the "realm of all realms," which is prior to any subject-object split. He asks: "Is this distinctive mark of the sort such that the adage asks about that which unspokenly also gathers into and retains by itself each thing-present; that which in terms of domain is not to be counted among men or gods and yet nevertheless in another sense is divine and human, plant and animal, mountain, sea and star?" (VA 278).[27] Although these things-present such as plants, animals, mountains, and so on, cannot be counted among the men or among the gods, these things are *in another sense* human and divine. These things-present are not objects radically separated from a subject; in some sense they are congruous with and indwelling to our existence. This, however, is a matter for later discussion.

The crucial thing here is that gods and men are defined out of their relationship to the illuminative clearing: men and gods are they who cannot remain concealed to that which never sinks away. Men and gods belong to the clearing and are congruous with it. In a sense they are instrumental in bringing about the perfective mode of the illuminative clearing, of bringing the latter into the fullness of its essence. Men and gods are enlightened (*erlichtet*); this means for Heidegger that they are: "implicated in and appropriated to the occurrent spectacle of the illuminative clearing, and therefore never concealed but rather *disensconced* . . . (that is) entrusted to the ensconcing illuminative clearing, which holds and restrains them" (VA 278–79).[28] Men and gods are not only illuminated but also assembled and held in presence by the illuminative clearing. They cannot exist except so contained; although they can turn away from the illuminative clearing, they can never escape it.

Man's susceptibility to the call of conscience, which can bring him back out of his state of everydayness and forgottenness, attests to the impossibility of remaining concealed to the illuminative clearing. Nevertheless it is only seldom that the call is recognized as such; usually its effect is merely to dislocate men from their familiar, secure surroundings so that men are merely the blind victims of restlessness and longing, weariness and dread, futility and suffering.

Aletheia, the occurrence of the illuminative clearing, is disensconcement, revealing/concealing. *Aletheia* is the unfolding/folding of the twofold; for the emergence of presence/things-present is also the retention, through ensconcement, of the simple fold of the twofold. Hence things can never be severed from their presence nor can presence be severed from things-present. Yet these are somehow brought out into a twoness, so to speak, which allows things to be distinguished as the unique entities they are and yet holds them ensconced in a togetherness of presence.

Aletheia, the unfolding/folding of the twofold, is never a one-way process of development from potentiality to actuality but always a two-way motion between the polar "moments" of the twofold: the simple foldedness which through ensconcement assembles and gathers and holds, and the emerging, illuminatory twoness of presence/things-present. Then does this mean that nothing ever "gets anywhere," that is, that we live in a semi-Eleatic world where progress and progression are impossible? Is the linear succession of days, months, years which constitutes human history only an illusion? How do we explain this? I think the answer lies in a note Heidegger made in a sketch entitled *Aus der Geschichte des Seins*. There he characterizes the incipient *aletheia* as "scarcely abiding and not returning back into the origin, but rather going forward into bare unconcealment" (2 Ni 458).[29]

Although the gods can supposedly see into the full essence of the illuminative clearing, men can only see that which appears to be revealed. The ensconcing character of *aletheia* shows itself only as mystery and forbidding bounded boundlessness; hence men turn away from the incipient, illuminative clearing and towards what they can readily grasp and see. Heidegger tells us: "Everyday opinion seeks truth in the plurality of continual novelty which is scattered out before it. It does not see the quiet radiance (the gold) of the Mystery which perpetually shines in the simple fold of the illuminative clearing" (VA 281).[30] He then quotes the Heracleitian fragment which says asses gather straw ere gold. Straw is shiny and tangible; it is easily gathered. Gold and its yellow radiance remain ensconced in the darkness of the earth; one must "dig up much earth to find a little gold." The full, incipient nature of *aletheia* forgotten, it degenerates into mere open, self-evident revealedness.

The consequences of this are staggering. We are still feeling them today. The forgottenness of the full nature of *aletheia* lies at the very heart of the metaphysical tradition which developed with the degeneration of *aletheia* into mere revealedness of things-that-are.[31] Heidegger sometimes refers to this as the oblivion or forgottenness of Being (*Seinsvergessenheit*) (WM 8–13; Holz 336). The forgetting of Being is not construed by Heidegger as an act of negligence for which we could rebuke various philosophers. Rather, the "going forward" of *aletheia* produces a drift in *aletheia* itself, a severance from the incipient simple fold.

The withdrawal of the simple fold and its mystery is so subtle that even the withdrawal conceals itself. Man is caught in the drift of the open realm—he becomes *adrift*, cut off from that which nourishes and anchors his essence. Thus man turns away without realizing it. He lapses into error in the epic, wandering sense, as one who is lost at sea. He wanders in and out among the things-that-are, caring for them, being overwhelmed by them, taking his nourishment and keep from them. This erring, wandering concern for the things-that-are while forgetting the open realm of their Being and the withdrawal implicit therein—this is what Heidegger calls "metaphysical." Thus both positive science and technology are according to Heidegger rooted in the metaphysical.

The history of man is intertwined with this concealment. History is the wandering course of man in the face of the withdrawal of the simple fold and its mystery. The history of philosophy can only be understood in light of this wandering course, this drift in the open realm. Conversely, the full significance of the drift and therefore the

Destiny and Ontological Difference 113

open realm itself can only be understood in its entirety if we consider its wandering course. Instead of a static realm or empty stage we have something more like an opening, a dynamic field which draws and withdraws. Thus there is a need to think back through the wandering course of thought itself, back to the origins of the problem of Being.[32]

Identity and Difference

Parmenides had spoken of *to auto* in connection with the unfolding of the twofold. Heidegger indicates that this *Rätselwort* somehow designates the Difference in its simplicity, as that by virtue of which the simple fold unfolds into the twoness of presence/things-present. Heracleitus had spoken of a hidden harmony between revealing and concealing such that every revealing is at once a concealing and vice versa. For him the unfolding (revealing) was also a folding (concealing). This incipient unfolding/folding withholds its identity as a simple fold, yet disensconces itself into the multiplicity of things. In their own ways both Parmenides and Heracleitus tried to characterize a fundamental, dynamic principle which is dual in character and yet somehow remains one principle. They are the precursors of the problem of identity and Difference.

The ancients in first raising the problem of Difference recognized the interdependency of sameness and difference. Without some kind of sameness only a radical otherness or dispersion could prevail. Vital to the notion of Difference is the provision that items be held somehow vis-à-vis one another and yet not be radically separated. The problem has always been to relate sameness and difference in such a way as to avoid self-contradiction: are sameness and difference "the same" or are they two "different" principles? Can one really attempt to define these two without not already presupposing one or the other or both? Or perhaps the very terms "sameness" and "difference" are inadequate because of their starkness, so that they force an exaggerated distillation of opposites?

At the outset of this inquiry the ontological difference was supposed to be a difference "between" Being and the things-that-are: "between" presence (*Anwesen, Anwesenheit*) and things-present (*das Anwesende*). This was found not to be any kind of objective distinction, because a mere distinction, "posited" by man, could not account for the fundamental way that man and the difference are related. The ontological difference has some "reality" of its own; perhaps in the end the ontological difference "is" reality.

The things-which-are cannot be static presences, for they must arise and possibly also depart within the open illuminative clearing. Moreover, Being itself cannot be static and eternal, for the character of the ontological lies not in an inert state of light and clarity but rather in an ongoing, an opening of the illuminative clearing. Thus Heidegger says: "Being must in fact already at the outset shine forth of its own accord and on its own terms, in order that any momentary entities can appear" (SvGr 111).[33] To consider the open realm as a static container which does not emerge and shine forth would be to suppress the full, incipient character of presence—and indeed to suppress not only the self-ensconcing but finally the emergent shining-forth itself: "Being abides, it is true, as *physis*, as self-disensconcement, as that which is evident of its own accord and on its terms, but there is also a self-ensconcement involved. If the ensconcement were lacking or ceased to be effective, how could a disensconcement take place?" (SvGr 113–14).[34] Instead of the dynamic, self-ensconcing emergency we would have a Platonic realm of already-present, static essences. Indeed, the failure to recognize the full, incipient character of the self-ensconcing, emergent illuminative clearing is at once the onset of Platonism: "Aletheia—scarcely abiding and not returning back into the origin but rather continuing forward into bare unconcealment—comes under the yoke of the Idea" (2 Ni 458).[35]

Platonism assumes that there is only an ambiguity at the apophantic level, the level of appearance of earthly particulars. Even though this particular object which I encounter as a chair is never literally a pure chair, evidenced by its transience and its dependence upon a context, there is a pure, eternal essence chair which is self-identical. Hence Plato spoke of likenesses or appearances, which have identities that are only relative and temporary, in contrast to the everlasting ideas or "whats" permanently present in unconcealment (2 Ni 218, 408–9; ZSf 16). Their unchanging whatness, their unconditional self-identity, is made manifest unambiguously to those who can engage in dialectic. During the historical course of philosophy it has become apparent however that Platonic Ideas are not at all permanently present. A set of ideas or "eternal truths" for one age or even one nation often dies with that age or nation. There is no eternal stability in the supersensible realm, and it is arbitrary to value the supersensible above the sensible. There may be ultimate structures and laws; but if so, there is no final interpretation of *what* these ultimates are. This is tantamount to saying that the essence of these ultimates is concealed: they may shine forth to man and illuminate things in our midst, but their shining remains highly ambiguous and cannot itself be captured as a "what."[36]

As Pöggeler points out, both self-identity and Difference can be encountered on many levels.[37] In a sense Platonism (as handed down through the metaphysical tradition) failed to recognize this. Most immediately there is the contingent, empirical identity which the objects around us seem to have up to a degree. It may be true that I encounter something *as* a rose, that *rose as rose* does not have enduring presence. Yet it would be equally mistaken to say that the thing I encounter *as* rose has no identity as rose whatsoever. Such a move would embrace a rampant phenomenalism which would deny the possibility of referring to any sort of stable objects in our midst. This would contradict the claim that Being ensconces itself in a plurality of things-that-are. In order to have a plurality we must have this and that as at least temporary identities. Yet rose as rose is only encounterable within the context of thing-that-is as rose. In other words, if something does not become present as a rose, there can be no sense to rose as rose. Yet what is it that becomes present as a rose? This is the riddle. We are at the point where identity comes to be haunted by the breath of Difference. Yet this is not to say that the something-to-become-present and the rose-present are definitely different things, so that there is a *Ding-an-sich* behind the rose. Rather, something comes to be present precisely *as* a rose. Ensconced in this "as" is the mysterious interplay of identity and Difference.

Heidegger tells us that "both Being and the things-that-are, each in its own way, appear forth from out of and on the basis of the Difference" (*ID* 61).[38] Stated otherwise, the twofold unfolds into presence/things-present, into be-ing/being(s). Prior to the duality of presence/things-present is a duality found within the unfolding Difference itself, namely the relation between the Difference (twofold) per se and presence/things-present, that is, between the twofold as simple fold and the twofold in its unfolded moment.

Since the twofold unfolds into presence/things-present, this means that the arising of the entities and the self-ensconcing emergence of presence are in a sense the same occurrence. This situation is described as follows: ". . . Being of what-is means: Being which the things-that-are is. The 'is' speaks transitively here, going-over. Being abides here in the guise of an overpass to the things-that-are. Being does not leaves its own place in going over to the things-that-are, as though they could be first without Being and then later made have to do with it. Being goes over (that) away, comes revealing over (that), which through such coming-over, first arrives as unconcealed on its own account. Arrival means: ensconce oneself in unconcealment: thus ensconced, continue to abide: to be the things-that-are" (*ID* 62).[39] To use a Hegelian term, Being "passes over" into the entities; Being abides as a disensconcing supervention (*entbergende Überkommnis*) such that entities are revealed. The supervention is not an actual coming-over which ontically leaves its source; rather the supervention is more of the character of an ecstatic projection or extension.

Water, for example, may be said to "come over" into the cup; but to do so the water must actually leave its source, for example, the pitcher. Because the water must leave the pitcher in coming over to the cup, the pitcher can someday be exhausted; then water will cease to be able to come over. The kind of coming-over involved here is nonecstatic; instead of a projection or extension there must be an ontic change of place.

An example of an ecstatic supervention might be seeing. To see something I must look "over there," beyond the ontic location of my physical body; on the other hand my glance does not actually leave my body. It is always I who is looking over there, and, while looking over there, I remain here.

Ecstatic supervention should be contrasted from the kind of supervention which prevails in the metaphysical tradition. The fact that what ecstatically supervenes does not actually leave its source does not mean that something else is sent instead, much as a king might remain ensconced in his castle and send out his messengers. Vital to the notion of an ecstatic supervention is that what supervenes is not estranged from its abode; this is in direct contrast to the situation of the messenger and the king, or, in Kantian terms, the situation of the *Ding-an-sich* and the phenomenon.

In describing the ecstatic supervention of Being, Heidegger says: "Being goes over (that) away, comes revealingly over (that), which only through such coming-over arrives as something-revealed on its own account" (*ID* 62).[40] The fact that Heidegger puts the *das* in parentheses indicates that the supervention is not thought of as a motion from something to something else, that is, to the things-that-are. The supervention does not act upon any kind of already-present substratum; yet through the supervention occurs the arrival of things-unconcealed. Being in supervening also disensconces.

Heidegger characterizes arrival as a self-ensconcement in unconcealment (*ID* 62). Yet our common sense tells us that when something arrives it is exposed, revealed, disensconced. Why does Heidegger use the terminology *to ensconce oneself in unconcealment?* Why does he not merely say unconceal or reveal oneself? Let us consider a few examples.

Let us say that the water leaves the pitcher and arrives in the cup. The pitcher-full-of-water, whereby water-in-the-cup is possible, does not ecstatically supervene. Instead the water becomes separated from the pitcher, so that water in the cup does not necessarily imply that the water came from the pitcher. There is no way for pitcher-full-of-water to ensconce itself in the arrival of water in the cup, and remain ensconced there.

On the other hand, a rough analogy to ecstatic supervention might be found in the way color per se supervenes, ensconcing itself in the arrival of blue, red, green, and so on. Blue, red, green are colors; but color per se does not appear. If color per se were to appear, it would have to be classed alongside blue, red and green. But then color per se would lose its self-identity, and there would be no common, regulative togetherness whereby red, blue and green appear as different colors alongside one another. These would not be recognized as colors at all and would become dispersed into a chaos of absurd singulars.

In a somewhat similar way we can say that Being abides as *disensconcing supervention* par excellence. Arrival occurs as self-ensconcement in unconcealment, whereby the things-that-are appear in their colorful multiplicity. The self-ensconcement expresses the fact that Being as ecstatic supervention does not leave its point of origin. Being maintains its integrity, keeps itself intact, while at the same time supervening and thereby providing the illuminative clearing, wherein the entities can arise and be the things they are. As such Being prevails as an ecstatic unity.

Supervention and arrival are not two separate things; nor can we say, strictly speaking, that one thing arrives and another supervenes. Heidegger tells us: "Being in the sense of the revealing supervention and the things-that-are as such in the sense of the self-ensconcing arrival abide, as the so differentiated, out of the self-same, the Difference" (*ID* 62).[41] When the Difference differentiates, that is, unfolds into presence/things-present, the latter arise in full color and multiplicity. Yet the things-present are not ontically separated from presence; indeed, so little is this the case, that the things-present are rather ensconced in presence, allowed to reside there. Were this not the case the things would fall into Anaximander's chaos of disjuncture.

The ensconcement of the things in presence signifies that presence and things-present, although differentiated and borne apart, nevertheless remain in a self-sameness (*to auto*, the simple fold of the twofold), much as red and blue remain in the self-sameness of color per se. The simple fold allows presence and things-present, that is, disensconcing supervention and self-ensconcing arrival, to be held apart while yet belonging to a kind of medium (*das Zwischen*, the Between). "The latter, the Difference as simple fold, first gives away and holds asunder the Between within which supervention and arrival are then held toward each other, borne-away-from-and-held-toward-each-other" (*ID* 62–63).[42]

By characterizing the Difference as the bearing-apart of supervention and arrival, rather than as a relation of otherness between the things-that-are and their Being, Heidegger tries to minimize the idea that two separate realms or things are involved. Supervention and arrival are in a sense the same occurrence; they are borne apart by virtue of the interplay of disensconcement and ensconcement rather than through any kind of ontic severance or specious otherness. The unfolding of the Difference prevails as the "disensconcing-ensconcing out-bearance" (*entbergend-bergender Austrag*) of supervention and arrival (*ID* 63).

In supervention the simple fold unfolds ecstatically as the shining-forth of the illuminative clearing. Were supervention and arrival not borne apart the simple fold would have to arrive as such, without any ensconcement. But to arrive is to ensconce oneself in unconcealment; there is no other way to participate in unconcealment except through ensconcing oneself therein. Hence if supervention and arrival are not borne apart there can be no arrival. Everything must vanish into endless dispersion. This may indeed be what happens in death: namely, where supervention arrives without ensconcement such arrival is made inaccessible to the rest of the world, that is, not allowed to participate in unconcealment. Thus the arrival of supervention involutes into a radical ensconcement which marks the very bounds of the illuminative clearing itself. Regarding death Heidegger says: "It [death] is as the most extreme possibility of human existence not the end of the possible but rather the highest mound (the assembling ensconcement) of the secret of the beckoning unconcealing" (*VA* 256).[43]

Because arrival involves ensconcement there can be no absolute, final arrival, no ultimate truth. Any kind of "progress" can only be relative, for it must consist not only in a successive revealing but also in a successive concealing. The revealing of the oak tree is the concealing of the acorn. Nevertheless Heidegger does not view the world as a nihilistic, endless succession of events, each to be superceded and concealed by the next. Rather he is saying that any kind of linear succession, nihilistic or otherwise, only appears to be what it is because we assume that each event arrives with such finality that it alone is viewed to be real; the given moment is viewed more "real" than what was or what is supposed to come.

Linear progression, strictly speaking, is not something totally linear but rather involves a reflexive interplay of self-sameness and difference. Were this not the case the world would consist only of infinite, inarticulate dispersion. Let us say for example that we begin with a set of conditions A and progress beyond it. This means that we leave A; to progress means that we continually move "further and further" away from A. But what gives us an awareness of "further and further"? We must continually compare our present position with respect to A; this requires a continual projection of our momentary self back to A but without a decisive return to A. Instead of returning to A, in progressing we involve the reflexive relation so that it turns around our momentary self. A comes to be regarded by us as a mere reference point with which we interpret our momentary position; hence the original, primary character of A becomes dissembled and distorted, and our momentary position has become an absolute reference point, back to which everything is reflected.

The fact that one can speak of a momentary self shows that the self is somehow drawn along, if not in a temporal succession, then in some kind of world course. Heidegger is not denying that there is temporal succession; he is merely saying that the notion of linear temporal succession is not sufficient to account for either the arrival of entities or the very nature of time itself. The concept of time as an objective, linear succession of events is objectionable to Heidegger for two reasons: (1) The "real" is thereby limited to the specious present, and (2) The fundamental role of time as that which gives and takes is suppressed.

The fact that history is possible shows that the past is not slipping away into radical concealment but rather remains in large part assembled in a common presence. Of course not every entity is so assembled and preserved. Paradoxically enough, the more completely something becomes ensconced in unconcealment the less chance there is of its historical preservation. The reason is that the more complete the ensconcement, the fewer traces there are of supervention. Finally an item's very unconcealment can be taken for granted, and the item does not even appear to have arrived. A Platonic Idea would be an example of this kind of entity. Other things are merely "there," such as rocks, soil, water. Others appear to arrive but not historically: chairs, tables, rain. Still others arrive with recurrent regularity: day, night, mealtime, sunrise, leaves, flowers. Yet why do we not call the arrival of these things "historical"?

Heidegger tells us: "Countless are the things in transition and those passed-away, seldom are those which have truly been, seldomer yet their preservance" (SvGr 107).[44] The historical is that which shapes the open realm and its drift; as such, the historical shapes human destiny. But why is this "seldom"? It is seldom almost by definition; the everyday is that which is open and evident on the apophantic level, that which presupposes the opening up of the illuminative clearing. The original opening up of the clearing itself is what is historically decisive for everything that can then reside in that clearing; this is the authentically historical, and it is seldom. What is truly historical is that which has been, not that which is by-gone, because the former, which occurs at the original hermeneutic level has already dictated the possibilities for particular comings and goings, as we have seen. What is historically so preserved is partially a matter of fate and partially left up to man. It is poetic man who writes down or brings to expression the opening-up of the illuminative clearing at the hermeneutic level. It is also man, historiographical man, who notes what particular things happen within the horizon of this clearing—happenings at the apophantic level. It is also man who discovers the old accounts of former ages. History thus involves a mutual constellation or harmonizing of man's attentiveness and the disensconcing supervention. Neither the mere blind release of events without anyone paying regard to them nor the mere watchfulness where nothing is granted unconcealment—neither alternative alone makes for a historical occurrence, perhaps not even for any occurrence at all.

Heidegger calls the constellation in which man and the disensconcing supervention are involved Er-eignis. Although this word usually means *event*, Heidegger here means something more fundamental. He tells us that he understands this word as eräugen, wherein the stem Auge, eye, is clearly discernible. Er-eignis thus means a letting-something-come-into-view, such that the seer and the seen are first defined out of a radically fundamental and singular occurrence (*ID* 29). Heidegger also connects *ereignen* and *sich eignen* (*to be adapted or made appropriate*). The latter word he understands in the sense defined within the poles of *sich zeigen* and *bezeichnen*: self-showing and designation (*USp* 258–59). *Ereignis*, self-showing-and-letting-be-seen, involves an appropriation or adaptation since it involves something coming forth as this or that. What comes forth as this or that must however always appear in light of a world which gives it a momentary identity. A jet plane could not come forth in the world in which Heraclitus lived, nor can a wellspring come forth as wellspring in our world.

A word in our language which captures some of the meaning of *Ereignis* is *spectacle*, that is, in this case "self-spectacle."[45] In the case of a spectacle the viewing and the viewed merge into a singular harmony. An appropriating is also implied, for the spectacular is allied with the unusual or remarkable, that which incites wonder. In order for something to incite wonder it must appear as wonder-inspiring. This too is relative to a world. In our world some things are appropriate for inciting wonder and others are not.

Self-spectacle characterizes the unfolding/folding of the simple fold whereby the illuminative clearing shines forth (*VA* 281). Self-spectacle is in a sense pure supervention reflected back into its simple unity and purity. Yet we know that this reflection does not dissolve into an empty identity; for supervention is *borne out* in two directions: in the reflection back into the simple fold and in the self-ensconcing arrival whereby entities are brought into unconcealment. "Being sends itself to man, in that as illuminative clearing, it furnishes for that-which-is as such a time-play-space. Being abides as that sent destiny, as self-disensconcing, which is at the same time a self-concealing" (*SvGr* 129–30).[46] Heidegger connects *Geschick* with *fatum*, which originally meant *oracle, that which is spoken* (*SvGr* 158). This ties up supervention and fate with the Heracleitian *logos* and the Parmenidean *phasis*, and all of these in turn with the primordial injunction which first calls forth the essence of man, mortal thinking and speaking (*WhD* 149; *VA* 242–45; *SvGr* 147, 158–61).

Finally we can say that self-spectacle underlies the nature of time. *Sein und Zeit* had characterized time as the final condition we appeal to, which gives or refuses the conditions under which we can act. Success is for us a matter of "knowing when," of seizing the opportunity. Our very congruence with the illuminative clearing is delivered over into an alteration of waking and sleeping, of light and darkness (SZ 404, 409–12). Regarding the ontological status of temporality Heidegger says: "Temporality 'is' not at all something-which-is. It is not, but rather brings itself to fruition" (SZ 328).[47] The words Zeit and zeitigen are related to our word tide as in tidings. Although Heidegger does not bring out the connection between time and the oracular nature of fate (Zeitung, New York Times, and so on) the connection should be obvious.

Because the self-spectacle, the supervention, the oracle, all involve a self-ensconcing, the actual arrival is borne apart. A harmony prevails between bearing-asunder and self-spectacle, which parallels that between the twoness of the unfolding twofold and the ensconcing/revealing simple fold. To summarize this relation between self-sameness and difference, between self-spectacle and out-bearance: "Nothing happens, the self-spectacle becomes spectacular. The beginning, on its own account—bearing asunder the illuminative clearing—takes leave" (2 Ni 485).[48] And with this departure began the drift of the open realm and the history of man: he who is on the threshold of the self-ensconcing disensconcing: he to whom Being "sends itself to us in that it withdraws itself" (SvGr 109).[49]

V

The Ontological Difference as the Realm of all Realms

Schmerzlos sind wir und haben fast
Die Sprache in der Fremde verloren.
(*Hölderlin*)

In the last chapter the ontological difference was explored as a principle of historical interpretation. Heidegger's main intention in the writings there considered was to reach into the origins of thought and show how the earliest thinkers experienced something of the unfolding power of thought and yet were unable to draw out the implications of these experiences. In particular, the earliest thinkers failed to locate the proper dimension of man and human thought. Throughout Heidegger's work there is the attempt to find this dimension, as we have seen; but the location of it requires him to do more than furnish interpretations of other philosophers—it requires that he engage in his own version of originative thinking. It should also be noted that in the course of Heidegger's thinking the ontological difference becomes less a principle, something abstract which conceptually organizes our thinking, and more and more a Region (*Gegend*) which commands our entire essence. This Region is characterized by him as a dynamic, twofold unfolding fold. If we ask about our relation to the ontological difference, the appropriate thing to say is that we are "in" it. The "in" is not a mere, physical, spatial designation here, but neither is it totally unspatial. Heidegger devoted a rather large section in *Sein und Zeit* to showing that physical or geometrical space is an abstraction founded upon and derivative from a more concrete, relational, dynamic space, the open realm. Human existence is spatial or ecstatic, being-in-the-world. All of our prepositions indicate this kind of ecstatic spatiality Heidegger is talking about. Some people claim that our ordinary speech is thus permeated with space "metaphors." But this would assume that abstract space is and ought to be the primary spatial phenomenon; this Heidegger challenges. In fact even mathematics could not do with mere geometrical space; for example the commutative law or any theorems dealing with the ordering of sets involve the same kind of spatializations that are found in Heidegger's talk about Regions (*Gegend, Bereich, Ortschaft*); neither can be said to be purely physical nor even purely abstract (geometrical) as far as I can see. Thus theories recognizing only abstract (geometrical) physical space cannot be adequate for the interpretation of Heidegger's Region, since they too are predicated upon a more fundamental openness.

It is important to see, also, on what level Heidegger attempts to conduct his own thinking. On the one hand, Heidegger is not talking about an object when he refers to the fundamental, twofold realm of all realms or the human Dimension; but neither is he engaging merely in conceptual analysis. The supposition that philosophers must do one or the other is predicated on a kind of Cartesian Dualism which must itself be overcome. Much of philosophy today is devoted, of course, to a clarification of traditional concepts and categories; certainly Heidegger too does his share of this. But if philosophy were to rest here, we would be tacitly admitting that we have all the categories and that all we need now are the fine points and clarifications. This Heidegger cannot accept. Philosophy must always continue its search for adequate categories since categories both reveal and conceal; and sometimes this requires that we be very daring, that we be thorough in our questioning of the foundations of philosophical thought. It requires that we have an open eye for the unusual, that we detach ourselves (*der Schritt zurück*) from what is commonly accepted, making a fresh attempt to perform a kind of reduction in order to gain access to that ultimate Region or "realm of all realms" which commands our essence (*Gel* 40; *VA* 272–73).

Gaining access to this Region is no simple matter, however. On the one hand, the reduction cannot be performed by means of a mere thought exercise such as Husserl had envisioned; nor can we simply refrain from conceptual analysis and expect this Region to make itself intuitively manifest as Bergson had thought. In this respect Heidegger's reduction differs rather radically from those of such philosophers as Bergson or Husserl, since they thought that reductions were relatively straightforward procedures. And insofar as they did not hunt for radically new ways to articulate and interpret their Regions, their reductions were only relative.

The Realm of all Realms 125

Heidegger's reduction involves an overcoming of the entire metaphysical perspective, not merely of certain aspects of it. But his reduction is not purely academic, just as his conception of metaphysics is not purely academic. Heidegger considers any open dealing with what-is in light of the category of being (Seiendheit) as metaphysical. Therefore the overcoming of metaphysics must involve a departure from the category of the "it-is." We must recognize that the "it-is" does not exhaust the realm of all realms, and in order to gain access to the latter we must "bracket" or step back from (der Schritt zurück) the former (ID 45-48).

Heidegger's historical interpretations and his analysis of the "negative" phenomena, dread, nothingness, death, conscience are attempts to prepare the way for such a reduction. In Was ist Metaphysik? Heidegger argued that the way to go beyond metaphysics is to confront the Nothing; but the forbidding character of the Nothing, expressed in the admonition that Nothingness is unthinkable, presents an unpassable barrier. To confront the Nothing seems thus to force the conclusion that there is nothing but the things-which-are, that is, that we must confine our concern to that-which-is. This is a typical reaction to our experience with the Nothing, according to Heidegger, because the Nothing is insidious. The immersion of man in the things about him is a direct effect of the Nothing upon him; however, the effect is rarely consciously experienced but for the most part dissembled into a vague restlessness and endless activity in the thing-and-people-world.

At this point we must however exercise great care in our own thinking in order to see the delicate point Heidegger is making. To say either that the Nothing is not worth bothering about because it cannot be thought or to say that the Nothing is worth bothering about because it can be felt or experienced—either of these alternatives assumes that the Nothing has some kind of objective status. But the Nothing is precisely that which does not stand over against us as a content to be found within our experience. This means that the key to understanding the Nothing must be sought by examining our experience itself. What are the experiences that have led men to speak of Nothing? How did they characterize these experiences? This is at least partly a historical question. Furthermore, the undissembled experience of the Nothing has been rare, and therefore we are forced to turn to the writings of those thinkers who attempted to express what this is. The early pre-Socratics, for example, had and wrote of such experiences; in fact it was Parmenides who finally concluded that the Nothing presented an impassable barrier and that man ought to confine his concern, both practical and theoretical to what is or can be and turn away from this void. Yet we must also ask, what led Parmenides to reject the Nothing? What led him to the conclusion that the way of the Nothing was impassable and unthinkable? What is this Nothing? Is Nothing the only alternative to what-is? What about Being (*Sein*)?

A recognition of the ontological difference allows us to see that there is something artificial about the impasse of the Nothing, that the alternative to what-is cannot be merely negatively contrasted with what-is. If we look at this alternative only from the standpoint of the things-that-are and the values rooted therein, of course it will appear nihilistic. The pre-Socratics failed to think out the significance of their experiences; they made use of the Nothing to give an account of what-is, but their eyes were riveted on the latter. They took the former for granted, concluding that the only way to think about something was to give an account of it, after the model of their thinking about what-is. It should come as no surprise that they would conclude sooner or later that the Nothing is unthinkable, that is, that it is impossible to give an account of it. But then how *could* we give an account of that which underlies and makes possible all account-giving and also all things-to-be-accounted-for? The attempt to do so would result in a fundamental confusion between that which grounds and that which is grounded. For Heidegger that which grounds in an ultimate rather than a merely ontic, relative sense can never be something-grounded. "Being 'is' in essence: ground. For that reason, Being can never itself have a ground which might give it foundation" (SvGr 93).[1] Thus even from the very beginning, pre-Socratic thought was only half-listening to the primordial call, although the pre-Socratics experienced a closeness to this call which has been subsequently lost in the history of Western thought.[2]

The problem of this Chapter is to explore the realm of all realms in which the Difference itself is rooted. If entrance to this realm is predicated on a departure from the "it-is," what will be man's deepest and most essential relation to the things around him? Must he literally renounce them? If we claim like Parmenides that they are ultimately unreal, the ontological difference loses its meaning. On the other hand, if they are real, then in what sense are we justified in departing or turning away from them? In many philosophies one could distinguish between what-is and a theory of what-is; we could say that we reject the *theory* that what-is (actuality) is all-important while still maintaining a kind of practical involvement with these things. Certain mystics or Platonists might maintain such a position. But to make a distinction between things and a theory about these things seems to imply that we can talk about things independently of our interpretations of them. This Heidegger would deny. Thus he is put in the difficult position of saying that the nature of things is somehow a function of our being-toward them or letting-them-be, while at the same time avoiding the conclusion that things are what we make them. There are no "things in themselves" for us to distinguish from our dealings with them; but neither can we reduce things to our dealings with them. What is the third alternative? It can only be seen by following through Heidegger's reduction, which involves not only an overcoming of the metaphysical perspective but also a developing of a language appropriate to the locality (Ortschaft) into which we are thereby ushered.

The usual, "metaphysical" way to look at things is as mere things, things-which-are. In doing so we invariably interpret ourselves also as mere things-which-are in a Region or ground of things-which are. This is what the metaphysical perspective does. The object of philosophy, under such a perspective, is simply to render a clear and comprehensive scheme of all the entities and their relations to one another. This means that philosophy loses its sensitivity to the problematic character of the Region itself. Heidegger suggests, on the other hand, that the Region is permeated with a twofold character; more correctly, the Region has the character of an unfolding fold. It is through and through dynamic, rather than being a kind of static, empty container of entities. Because this is so, the Region ought to continue to always be a challenge to the philosophic thinker who dares face up to it. Now, entities themselves were argued to be ensconced termini of the disguised withdrawal of the fold itself. Finally, Heidegger argues that man ought to be viewed as belonging to the ontological dimension of the unfolding fold (insofar as he thinks) and also to the ontic dimension of the entities (insofar as he exists on a day-to-day basis with other things-which-are and is subject to the same laws and fates that they are).

To put it another way, Heidegger thinks it would be a drastic mistake to claim that Reality were exhausted by the totality of what-is. The phenomena of conscience, dread, guilt, death all take us beyond what-is into the murky regions of ensconcement and withdrawal. These cannot be regarded as mere psychological phenomena, as we have already seen. Therefore the presence of these phenomena is an indication that the dynamics of withdrawal and ensconcement are as much a part of our Region as the so-called entities. The very idea of simple entities is a mistake; these must be conceived dynamically, in proper polar opposition and harmony with the complex Region of supervention/withdrawal/ensconcement.

Because the entire realm of what concerns us includes more than the realm of the so-called entities, defining the former rather than the latter as real does not confine Heidegger to an arbitrary subjectivism but actually does the opposite. We are thereby allowed to leap away from the "it-is" and to recognize a more fundamental character of the realm in which we abide. This latter "realm of all realms" (VA 272–73; Gel 40ff) Heidegger calls sometimes das Ereignis, spectacle-on-going, or Gegend, Polar-Region (ID; Gel). Sometimes he refers to it as the unfolding fold in its simple foldedness, the Heracleitean realm of that which never sinks away. One thing to see here is that Heidegger is not invoking a space-metaphor in speaking of Regions or realms of all realms as much as he is attempting to characterize a dynamic locus of the ultimate and most important ongoing; his Region is "spatial" only in the way that a vector for acceleration is spatial. But the Region is ecstatic in much the same way a vector is ecstatic, and it is on-going or event-ful.

Heidegger's reduction takes the form of a leap (Sprung) or a step back (Schritt zurück).[3] This reduction cannot be undergone on a theoretical level as some kind of thought-exercise, but involves a total reorientation. We saw earlier that what we must leap away from or step back out of is the tyranny of the "it-is," the tyranny under which the entities become the standard for interpreting all human experience. But now Heidegger is speaking of an even more radical leap; we must leap away from what-is into that free openness, the Region of all Regions, which lodges not man as an ontic existence but rather man's essence (Wesen). Yet man as thinker must not merely use the primordial summons to interpret himself, as though his existence were its own problematic essence and everything centered around it; he must complete the circle and use himself to attend properly to the primordial call. Here is where the radical leap is required, since man must become acquiescent (gelassen) and maintain an open receptiveness to the call and the withdrawal therein implied (Gel 44–54). Only in this way can the metaphysical impasse be overcome.

The departure from the "it-is" looks, from the standpoint of metaphysics, like a jump into the chasm of nothingness. To conclude this is to reason somewhat hastily however; for there is no reason to presume that the sole alternative to what-is must be radical, absolute Nothingness (2 Ni 52—54). Because of the forbidding character of Nothingness, an attempt to leap nihilistically immerses us all the more in the realm of what-is and hence is not really a leap at all (WM 22). We must remember that the free openness is far from empty. It is the fateful point of intersection of earth and world and as such contains all the traces and tracks of the unfolding/folding (Holz 46–65). The leap is a step back in the sense that we step into the Region in which we already are: the ecstatic moment, self-spectacle, in which what was, is and will be is all folded and unfolded together.

Heidegger calls Being *die Leere und der Reichtum* (2 Ni 247). On the one hand Being is the illuminative, open realm of presence, but on the other hand Being is the forbidding, empty withdrawal of the open realm. Being is on the one hand all-pervasive (for all things are) and on the other hand radically singular and self-same with itself. Being is on the one hand the most certain and most fundamental principle whereby entities are allowed to arise in their intelligibility and on the other hand a chasm (2 Ni 251–52).

In making the leap we must acknowledge both sides, the coming-over and the withdrawal. The leap must affirm that into which it leaps; such a leap is far from being nihilistic. In making the leap we recognize our radical finitude; we recognize that all things—including ourselves—belong to the realm of the twofold rather than to us. We recognize that our very essence has been summoned forth, that we are not self-justifying or *freischwebend*.

In leaping we do not leave the entities behind; rather they take on a new light for us (SvGr 119). As the ensconced termini of arrival they become tangible bearers of the proximity of the self-ensconcing supervention. Instead of losing their integrity altogether, the entities acquire what for Heidegger is a new and more fundamental integrity, one measured by the "realm of all realms" rather than by categorical standards.

Measure and Dimension

When we come to a new place, we immediately try to orient ourselves, to find out where we are. To do this we make use of certain measures, certain relationships (two blocks from the bar, east of the park, and so on) in light of which we interpret our own position.

What we take as our measures and categories of interpretation are a function of the character of the place we are in, provided that these are familiar enough so that we can appropriate them to our own way of life. We must explore and look for those familiar things that will make us feel at home in the new place. To feel at home is to understand where we are, to have a sense of orientation. To feel strange is to be immersed in a context which we are unable to measure and comfortably interpret. We don't know what standards to apply. Because of the interdependence of standards of measure and place or contexts Heidegger tells us: ". . . standard and area to be measured are not two distinct or even separated things, but rather one and the same. The standard provides and opens an area in which the standard is at home and can be that which it is" (GrD 35).[4]

Heidegger considers the problem of measure to be fundamental to philosophical thinking because it is fundamental to the human realm, whether this realm is superficially or authentically dwelt in. The problem is to come up with adequate standards and categories whereby man can orient himself and interpret what he most truly is. This must be done in light of the Region in which he is, and only when this Region is properly measured can man be said to dwell or be at home in it. But what kind of a measure can measure a Region which has no Region beyond it and which ensconces and withdraws itself?

For Heidegger the problem of what standard to use cannot be dealt with separately from the problem of how measuring takes place. Measuring standards cannot be formulated abstractly; in this sense Heidegger is an operationalist. Thus the problem of measure develops into two questions for Heidegger: (1) what is it fundamentally to measure, and (2) how can that measure occur which most accurately discloses the nature of man and his abode?

When we speak of measure we usually mean a comparing of two things, for example, a measuring of a distance between two points by means of a rule. But this measuring-by is not the most basic form of measure, because it has already presupposed a measure in the form of a rule. In making use of measure, measuring-by does not let the true character of measure itself come to light; nor does measuring-by make clear the relation between spatial or temporal distances and measure per se (USp 209).

Comparative measuring, that is, measuring one thing by another thing, is easily recognized as an aspect of calculative thinking. To calculate is to determine, to compare one thing with another, to weigh one thing off against another (SvGr 168). To be able to compare two things or to weigh one off against another presupposes that one of these things is determinate; this already-determinateness of something we call quantity. Comparative measuring is quantitative, and numerical calculation is only one kind of quantitative measuring. The words *quantity* and *measure* are often used interchangeably, although Heidegger does not deal with this point. A quantity of grain, for example, is measured out. This means two things: (a) some grain is apportioned off from the rest, and (b) some grain is limited, that is, boundaries are fixed to it.

If we know in advance that we want to use all of the grain in the granary, we will not bother to measure it out. It is the fact that we are interested in *some* grain rather than in all of the grain which gives rise to the measuring. However our interest in "some" is not an interest in something abstract and theoretical. The grain has supposedly been separated off for a reason; it will be appropriated for something. Boundaries have been fixed to "some grain" in order that the amount of grain be appropriate to the purpose it is to serve. A handful will not plant a field, nor is a full granary necessary to bake a loaf of bread.

We are now approaching a more fundamental meaning of "to measure." On the one hand there is a distribution or apportionment; on the other hand there is a limitation, a moderation. These two are interdependent. To apportion is to fix limits which separate a part off from the whole; yet the ability to fix limits presupposes that what is to be limited is already determinate enough to be a possible locus of boundaries and therefore apportionable off from the whole. These two also presuppose some standard, purpose or value which we take as our measure, for example, grain to plant a field or grain to bake a loaf. This means that at the bottom of all measuring is a *measure-taking*, an apprehension of the standard with which we can measure something and thereby limit it and apportion it (VA 196). Yet what is the source of this measure-taking?

The original source of man's measure cannot be purely ontic or based on what is within the world, since the ontic is itself derivative and measured in light of the world in which it is disclosed. Yet the ontic cannot be ignored insofar as much of man's Region is disclosed in terms of his relations to the things-which-are (*die Seienden*). This implies that the measure-taking cannot be merely ontological either. Could the source of measure-taking be ontic-ontological? We have seen that the ontic-ontological disclosed only one side of the Region, namely the side turned toward us. We have also seen that man's essence cannot be illuminated unless his full relation to the ontological difference is spelled out, including the self-ensconcement (withdrawal) as well as the self-illumination (supervention). The measure must itself be twofold: ensconcing/illuminating. But let us examine the problem from an operational point of view.

When we measure we must ecstatically traverse the open realm; for only by doing so can we fix boundaries, experience the way in which something is apportioned off or apportion it off ourselves. In ecstatically traversing the open realm we traverse the span between our starting point and the very boundary of the things-that-are; when we have reached this boundary point we return. The return is necessary if there is going to be a true measurement. If we do not return there are two possibilities: either we keep on going or we simply stop, riveted on the boundary. An example of the first might be some sort of schizophrenia, whereas the second might be illustrated by an animal consciousness which lacks a historical-factual sense. In the first case no boundary has been set and hence no measurement has taken place; in the second case we ignore our starting point and hence do not apportion off the span which we have traversed.

This shows that what is measured, strictly speaking, is what is *between* the boundaries fixed, rather than the boundaries themselves (VA 196). This does not yet mean that the open realm is a quantitative, objective space however; it only means that boundaries can only be fixed by ecstatically traversing what is between them. There must be at least two boundaries fixed to constitute a measurement; for only if there are two boundaries can there be an apportioning off. Yet it is important to note that only one of these boundaries need be vivid and discrete, namely our starting point. The other boundary need only compel us to return; if we are compelled to return we can speak of a measuring, albeit one which is not numerically quantitative in character.

The span bounded by our point of departure and our point of return Heidegger calls *Dimension*. Dimension is the full span between the ultimate bounds of human existence: the free, open realm of the illuminative clearing: the self-spectacular realm of all realms. Dimension is the realm allotted to us, bounded by earth and heaven (VA 195). Man dwells *upon* the earth and *beneath* the heavens (VA 149). As such, man oscillates, looking upward (*aufschauen*) and traversing the Dimension; this constitutes the very essence of man as he who dwells. "Man's dwelling rests in looking upward and surveying the Dimension in which the heaven belongs just as much as the earth"

(VA 195).[5] Ideally, man does not survey his Dimension merely in earthly terms; nor does he survey it totally in terms of the beyond. As the being in between, his measure-taking or surveying must involve both heaven and earth. Yet even these two together do not suffice; they do not furnish the measure which is needed for the survey. What is this measure and where could it come from?

Before this question can be answered it is necessary to note what kind of surveillant traversal is going on here. What is there to be measured? What is to be gained by such a measure? The surveillant traversal itself corresponds to the dynamics of authentic selfhood. We are at once involved in an ecstatic extension of ourselves out to the very limits of the open realm and in a return to our factical situation. In returning we encounter the things-that-are (SZ 308, 365–66). Thus measure-taking and surveillant traversal is not something which we can choose to do or not to do. We are always at any time undergoing this traversal and always at any time doing it in light of some measure which has been taken. However, if the measure is inadequate, the traversals which go on in light of it will fall short of the full essence of man.

The same dynamics are expressed when we are said by Heidegger to be drawn along the open realm by the withdrawal of the simple fold. Although drawn along through the open realm, we can only follow to the limits of the illuminative clearing; we cannot enter into the realm of ensconcement as such. We are compelled to return, and in doing so we are placed anew in the midst of the things-that-are, which are also the revealed termini of the self-ensconcing withdrawal. Thus our return to the things-that-are and the withdrawal of the unfolding fold are one and the same.

According to Heidegger it is the poet who has the closest relation with the Mystery, for it is he who is open to the hermeneutic level of originative disclosure. The poet preserves the mystery of self-ensconcement as a definite ongoing. He does not try to compel its secrets as does the scientist. The age of the poet is the age in which the divine messages are self-ensconcing. Heidegger calls it the time of impoverishment (*dürftige Zeit*) (Holz 248). This is the age of the ontological difference, of the mysterious self-ensconcing simple fold which nevertheless clothes itself in the things familiar to man.

The poet gains a closeness with the self-ensconcing simple fold because he does not have a prejudice totally in the direction of the "it is." The self-ensconcing fold is mysterious and holy. It is holy because it "heals" (*heilen*) or makes whole the multifarious fragmented character of the things-that-are. The holy bounds the limits of the illuminated realm, thereby first allotting the realm of the known to us. The holy is the source of disclosure at the hermeneutic level. Thus Heidegger speaks of the unknown divinity who, although remaining ensconced, sends a message of this self-ensconcing withdrawal. This message arrives in the form of the open-ended boundary, the bounded boundlessness of the heaven. The very openness of the heaven, which directs our vision away from the familiar reality of the earth and yet also bounds and encompasses this familiar reality, suggests the mysterious and the divine: "Thus the unknown god appears as the unknown one per se through the openness of the heaven. This appearing is the measure according to and by which man measures himself" (VA 197).[6]

Note that Heidegger does not say that man measures himself by a god. In fact he explicitly denies this (VA 197). There is no divine revelation to tell man what and who he is and how far his scope extends. Yet there is a measure of man furnished in the way the god, while remaining unknown, can still indirectly manifest itself in the openness of the heavens. The measure is not a "given" which is accessible to just anybody; it cannot be taken or grasped at the apophantic level. It is only the poet who remains open to the unknown and lets it be in its ensconcement who is able to take such a measure. The poet, to take the measure, must heed and then transmit.

The case of the poet deserves to be contrasted explicitly with ordinary ontic cases involving storage and transmission of information. There may be a sports event playing in a city such that many sport fans cannot attend it. There are TV cameras and radio equipment on the scene however so that such fans can watch or listen to a description of the game on their TV or radio. No real problem of interpretation is really involved here; even the describer of the event describes in accordance with preestablished rules that define the significant moves and plays and outcomes in the game. The meaning of the event does not really add new possibilities to our Dimension; hence it does not really involve a deepening of the Dimension in the way that poetic experiences do. The message of the event is clear such that anyone who is on the spot or anyone who watches or listens as the message is transmitted knows what happened. Any problems of meaning in this case for example, disagreements as to what the event means for the careers of the various players involve the apophantic level only. There is no urgent problem of revealing/concealing for sports analysts.

Now the poet is able to notice what the nonpoet does not notice but for different reasons than is the case for someone who has to work when the game is being played. Is the poet then like the geometer, who can notice certain relationships in a figure that the nongeometer simply takes to be a design? Again, a contrast must be drawn. The geometer differs from the nongeometer in that he has a knowledge of certain rules which constitute the body of knowledge called geometry. Learning geometry is a matter of mastering these rules and using them to discover other rules. Here again the problem of meaning is essentially an apophantic one, although the presence of "undecidable formulas" suggests that there is a hermeneutic problem for mathematics. Nevertheless, most mathematicians avoid it and behave as though there were no such problem. In the case of the poet and the nonpoet, however, it is not a matter of having failed to master rules, but more a matter of fate and orientation. The poet is gifted, singled out. Becoming a poet is not a matter of learning rules. Virtuosity with language is not poetry.

The problem for the poet is to receive the measure given in the appearance of the unknown god. He must name it, call it forth so that men may apprehend it; but at the same time the poet must preserve the intactness of the Mystery. The message has no familiar content although it is clothed in what is familiar to man, namely the visages of heaven. "That which remains alien to the deity, the visages of heaven, this is what is familiar to man. And what is this? Everything which in the heaven and consequently under the heaven and consequently upon the earth glistens and blossoms, resounds and emits fragrance, rises and comes, but also goes and falls, but also laments and remains silent, but also fades and darkens. The unknown sends itself forth (in the guise of) what is to man familiar but to the deity alien, in order to remain therein ensconced and protected as the unknown" (VA 200).[7] By summoning forth in the visages of heaven that which ensconces itself, the poet thereby calls forth the strange element which lurks in the familiar things around us. The invisible, the unknown, is summoned forth and yet allowed to abide in its mystery. The poet speaks in pictures, thus allowing that which ensconces itself to be seen without jeopardizing its integrity. "The poetic saying of the pictures assembles brilliance and resonance of the heavenly phenomena together with the darkness and silence of the strange" (VA 201).[8]

The heaven itself contains the brilliance of its altitude but also the darkness of endless expansion. The heaven is akin to the Heracleitean fire which assembles and brings forth (those things) such that the things are allowed to arise and to persist in their familiarity. Yet, also like the Heraclitean fire, the heaven not only gives measure and intelligibility of things but also can withdraw these and enshroud us with the darkness of night. Heaven is the abode of the unknown god; but as such heaven is not a supersensible realm or the promised kingdom. Man is only man qua dwelling *beneath* the heaven and *upon* the earth.

From the heaven comes light and warmth which allow the things to be illuminated and to thrive upon the earth. Thus the heaven is something highly dynamic in character; it is not merely an abode or static realm or empty space. The heaven gives us a sense of continuity and orientation: "The heaven is the arching sunpath, the vacillating mooncourse, the wandering lustre of the stars, the times of the year and their solstices, light and twilight of the day, darkness and brilliance of the night, the hospitable and unhospitable character of the weather, flight of the clouds and bluing deepness of the aether" (VA 150).[9] "The lustre of the heaven is rise and fall of the twilight which holds all which is revealable. This heaven is the measure" (VA 201).[10] The heaven assembles and holds in proportion the known and the unknown, the disensconcing and the ensconcing, the divine and the familiar. The heaven, interpreted in this special sense, provides therefore the measure which the poet takes to fathom the human Dimension.

To leap away from the "it is" is to traverse the entire Dimension of man, bounded by the familiar earth and the boundless boundedness of the open heavens. In leaping we approach the realm of the poet; we dwell in his neighborhood (USp 179). It is the poet who bears the positive relation to the mysterious; he guards the divine character of the unknown and does not try to turn the unknown into the known, as does scientific man.

Yet this leap should not be construed as a leap away from the earth; indeed Heidegger is of the opinion that scientific man pays as little regard to the earth as he does to the holy. He says regarding calculative thought: "This kind of thought is about to dispense with the earth as earth" (USp 190).[11] The reason is that science tends to make the earth just another planet; man investigates the earth as an object and ignores his own facticity, his own earth-boundness. Earth is for Heidegger most fundamentally ". . . that which serves and bears, the blossoming fruition, extended in stones and waters, emerging into vegetation and animal life" (VA 149).[12] The earth bears, supports and sustains the things with which we are familiar. These things are so familiar that we take them as complete units in themselves and forget that they have been granted, supported and sustained by the earth. For example, the things are heavy; they all fall toward the earth. But we tend to look at this fact as a scientific law concerning the behavior of bodies rather than as reflecting the earthboundness of the things around us and ourselves. We *abstract* the familiar things from their context which bears, supports and sustains them; we give up the earth qua earth and thereby constrict the earth-directed boundary of the human Dimension. Hence the leap must not only reach out into the heaven, but it must also reach down into the earth, recognizing the latter as an equally important aspect of the human Dimension.

Whether or not the poet is allowed to traverse the full range of the Dimension, that is, whether or not he is allowed to take the measure and fathom the Dimension with full profundity, is not left in his own hands. The poet cannot act alone nor can he merely create; rather he must heed and wait. It is the ensconced gods who bring the message of the mysterious to the mortal poet—the message of the simple fold which in unfolding/folding grants and preserves (*USp* 143).

When man imposes explanations upon the mysterious the latter disappears on principle; even though the explanations be inadequate, the gods withdraw their message and man rivets himself on the familiar things within a shrunken horizon. Under such circumstances the poetic is stifled and only the empirical and utilitarian thrive; a nihilistic age sets in. Yet Heidegger does not view the withdrawal of the message as a result of man's impiety; rather man's turning away from the Mystery occurs together with this withdrawal.

The poet can only receive the measure by which to traverse the Dimension if the gods are graciously disposed. If the gods allow it and if the poet is heedful, then the message arrives. Neither the mere heeding of the poet nor the mere grace of the gods is sufficient in itself (*VA* 204).

The departure from the tyranny of the "it-is" allows man to traverse the full scale of his Dimension: the "Between" in which man dwells, *upon* the earth, *beneath* the heavens and *before* the gods (*VA* 195, 149). As such the Dimension takes on the character of a locus of the *foursome*: gods, mortals, earth, heaven (*VA* 150).

In the essay *Zur Seinsfrage* Heidegger writes the word *Sein* with a cross through it. The cross does not only indicate that being is not an entity, that is, that Sein does not stand for a thing, but also that the notion of Being must yield to the notion of a dynamic locus of intersection of the foursome. Only then can we rid Being of its quasi-objective character vis-à-vis man (ZSf 31). Being must yield to the Dimension, that realm of all realms or Region whose measure is not something empirical or metaphysical but is rather the underlying harmony between ensconcement and disensconcement (VA 199–200).

World and thing

Throughout his work Heidegger has emphasized that man's most primary activity is not one of representing objects but rather surveying his Dimension—looking up into the heavens and back to the familiar earth. Thus, man's existence is ecstatic; instead of looking at man as some kind of static focal point of representations, Heidegger looks at human existence as a kind of field or topos of dynamic traversals. Man traverses the open realm; but this is not done instantaneously. What is traversable must be pavable; the mission of human thinking is to build paths along which to traverse the open realm, the Dimension, in response to the primordial summons.

Most of the paths we pave do not get us into the heart of this Region, however, but at best use the Region only as a kind of roadbed. This is to say, most of our thinking is not oriented toward the Region itself but rather towards the things-which-are, which the Region allows to be present. Because the presence of these things-which-are is often so obtrusive and demanding, man becomes preoccupied with the task of ordering these things and giving an account of them, and the Region itself fades from the sphere of human attention. This is, according to Heidegger, the crux of metaphysical endeavor; it does not attend to the Region, but only to the things which issue forth.

Gaining entrance to the Region itself in any full sense involves a departure from the "it-is," a kind of "reduction." But what kind of departure is this? What ought to be the relation between man and things-which-are? In his dialogue between a Japanese and an inquirer, Heidegger tells us that he thinks of the departure as an arrival of what-has-been (Ankunft des Gewesenen) (USp 154). Elsewhere he says: "But the true Time (Epoch) is the arrival of what-has-been. This is not the by-gone but rather the gatherage of that residual presence which precedes all arrival in so far as it ensconces itself as that gatherage back in what it was at any time earlier" (*USp* 57).[13] Again he says: "... what-has-been. By that we mean the gatherage of that which does not pass away but rather dwells, that is, lingers, in that it grants new insights to remembrance" (*SvGr* 107).[14] It should be noticed that *das Gewesene* is not a *something* which has been; that would have to be at this or that time and not some other time, and thus its departure would be such that when it is gone it is by-gone. A sirloin steak would be a supreme example. But *das Gewesene* is not this or that thing-gathered but rather the gatherage itself—a kind of dynamic reservoir which holds all (significant) things-which-linger (*währen*), be they present or past. In fact in a sense this reservoir also holds all things-that-will-be, although this is ensconced from the view of mortals. Thus the past and the future have ontological status for Heidegger, although mortals have no access to the future. The past however can take on a character of living immanence (as opposed to transcendence) through the departure from what-is. The boundary between what-is and what-was, between past and present, must be overcome.

The departure from the "it-is" also involves a shift of concern away from viewing things in isolation and towards their problematic unfolding/folding context of presence, their world. For Heidegger there are no "things-in-themselves" which could be determined independently of the human standpoint and their context of presence. This is not to deny that physics and mathematics make investigations into what-is. But these too are thoroughly human, and besides, they do not really deal with *things*; they attempt to deal with what-is independently of the particular world in which it is given. For Heidegger, however, things *are*, relative to a world which discloses (and also ensconces) them. If a world is transitory and not long-lasting, the things in it will also fade away. Thus certain styles of dress or certain implements may be mere curiosities in our world. Items such as the forge-bellows, which meant a great deal to people in former ages, are mere curiosities for our world. Jet planes, thermos bottles, ball point pens are items which belong typically to our world. Even the world of Newtonian particle dynamics is a historically disclosed world; it was pervasive enough to dominate Kant's philosophy, whereas it had no hold on Aristotle.

The question arises, is there one super-world which all particular worlds have in common? Is there an essence, world per se? In a sense Heidegger answers this question positively, for there is the Region, the realm of all realms. But this Region cannot be arrived at by abstracting what is common to all particular worlds. The Region is self-ensconcing; we cannot arrive at it through an in-duction but only through a re-duction, a stepping backward. The Region is not an essence in the sense of a universal which can be logically arrived at. For induction can only give us what is commonly revealed, never what is ensconced. Thus any particular world always has world per se *ensconced* within it.

We can only determine what a thing is if its world is predisclosed. Yet in our everyday rapport with things the world remains very much in the background; worse yet, our familiarity with the bounds of the human Dimension is at best intuitive or subliminal. To understand our world, however, we must let world or the human Dimension per se arrive. To let world arrive, be present in its full significance, requires a double stepping back: out of the everyday and out of metaphysics (the assent to it-is). Both have in common a "sense of Reality" which takes its measure from what-is.

We can never find the full meaning of things around us by just looking at them. This is the mistake of empiricism. The Dimension is closed to the empiricist because he will not look beyond the things lying around us. To apprehend an item we must apprehend it in light of something which ensconces it and allows it to be present— some kind of predisclosed "categories" we use. But where would these "categories" come from? Heidegger rejects the view that categories are mere mental constructions, since, were this the case, our categories would be useless except for talk about mental phenomena. Nor can categories be arrived at inductively, for as Kant pointed out, they would then lose their significance as disclosers and would instead be part of the disclosed. Besides, we must perform inductions in light of something. Thus any categories arrived at by induction would already presuppose categories.

Categories are disclosed through the surveying of the Dimension; only by looking towards the boundaries of human existence—a kind of speculation—can we arrive at adequate categories, which do not rigidly try to fix things but rather *invite* their presence. Only by such a traversal of the human Dimension are the things around us allowed to arrive in their full significance—not simply as things lying around us but as things which reflect a world within them. *World* here is the dynamic, reflective interplay of the foursome, the men and gods, heaven and earth (VA 178).[15] The term *to-be-in-the-world* had characterized the relation between man and world such that world was not an objective structure but rather an *existenzial* structure (SZ 88). However, *existenzial* is now broadened in meaning to explicitly include more than the human element because world as *existenzial*, although not an objective structure, is the locus of the foursome rather than of human existence alone. World is at once human and divine, earthly and heavenly. Only within this interplay can the things within the world be properly understood and interpreted.

Thus, too, we are not to think of the foursome as four bodies alongside one another; rather, they are elements of an interplay, whereby each reflects itself into the others and reflects back the others into itself. The interplay of the foursome is described by Heidegger as the on-going, self-spectacular reflective interplay (*das ereignende Spiegel-Spiel*) in which *keines der Vier versteift sich auf sein gesondertes Besonderes* (VA 178). The interplay is reflective and fluid, so that none of the four solidifies itself in its own area. Through looking at and being a part of this reflective interplay, man receives the categories or measures within which to interpret the things in his world.

Because man is only a part of this fourfold interplay it is impossible for human reason to achieve a "point of neutrality" that would allow us to fathom the essence of world per se. World as the on-going, self-spectacular reflective interplay remains inexplicable, without reason or ground: "As soon as human cognition demands an explanation, it does not transcend the essence of world but rather falls beneath, under the essence of world. The human attempt to explain does not at all reach into the simplicity of the simple fold of dynamic world" (VA 178).[16] Human reason cannot penetrate the simple unity of the fourfold interplay; nevertheless we do experience this unity, as I shall explain shortly.

Heidegger characterizes the simple unity of the fourfold interplay; as a "round dance" (*der Reigen des Ereignens*). A round dance is a ring which rings, that is, a configuration which turns as a whole while its constituents are also turning, as is the case with the solar system (VA 179). The ringing motions of the round dance are assembled into the *Gering*. Here Heidegger is using the prefix *Ge-* to indicate an assemblage of ringing motions, wherein each of the four bends back into its own zone, while at the same time acquiescing into the onefold foursome (VA 179). The fourfold interplay is also a onefold foursome. The latter term stresses the acquiescent, yielding character of the four, while the former term stresses the fact that there are four unique constituents or moments of the interplay.

The *Gering* appears to involve two dynamic moments, namely, the onefold foursome and the fourfold interplay. In the former case, a simple unity prevails, whereas the latter emphasizes an ecstatic, dynamic Dimension involving four participant constituents. The latter has been characterized as world. The simple, unfathomable unity appears to man in the guise of a thing. The two moments of the twofold *Gering* are world and thing.

The thing is said to allow the foursome to linger (*Verweilen*), confining the participant constituents into a simple unity. Hence the thing assembles the four into their locus of interplay; the thing assembles and brings about world. On the other hand, the thing itself is a simple unity forged out of the confining circle of the *Gering*. The latter compels the four into a simple togetherness; the thing stays or restrains this simple togetherness so that it does not disappear into radical ensconcement (VA 179). The thing thus preserves the simple fold or unity of the foursome.

The example of a thing used in "Das Ding" (VA 163–81) is that of an earthenware vessel (*Krug*), a pitcher. In "Bauen Wohnen Denken" (VA 145–62) he uses the example of a bridge. Both of these examples would appear to be chosen with great care. They both have "ecstatic" functions which bring together and then disperse or let flow apart. To be able to assemble the foursome, the thing must somehow "reach out" while at the same time remaining localized within itself (*Insichstehen*). The bridge, precisely by being localized where it is, assembles earth by bringing the banks of the stream together as a single landscape. It withstands the various kinds of weather which the heaven sends. Men go over the bridge to come together and to take leave of one another. The bridge hovers over the stream and thus localizes in a concrete way man's hovering attempt to overcome that which trivializes and fragments him in order that his life be made whole. Thus the bridge gathers together before the gods, hidden though they be; in this sense the bridge lets the Holy be present, that which can make us whole (VA 152–53). Heidegger's analysis of the pitcher runs along similar lines. While remaining localized within itself, the pitcher can offer out (*schenken*) a drink of water or wine. Both water and wine involve the "marriage" of earth and heaven: the rains which nourish the wellspring or the vines. Drinks can be offered by man to man. But this empties the pitcher. The offering-out of the drink can only occur when the pitcher has something to offer. This alludes once again to the ominous, hidden gods and to the sudden reversals of want and plenty which befall man in hidden, unexpected ways. Thus the emptiness of the empty pitcher means much more than space or air molecules; it bears witness to a time of want, when the favor (*Gunst*) of the Holy is turned away (VA 168–72). A thing is a thing to the extent that it assembles the foursome. This means that not all objects lying around us are capable of becoming things. Could razor blades, cigarette butts, lipstick or popcorn be regarded as things in the Heideggerian sense without leading to a ludicrous analysis? To assert that such items reflect and assemble the foursome yields an embarrassing problem similar to the one Plato faced concerning how many Forms there are and whether there are Forms of mud and nails.

Is this a weak spot in Heidegger's thought, namely, that he does not account for all objects but only for the ones he calls things? I would say that it is no more a weak spot than is any distinction between necessary and accidental being. An attempt to have an ontology of accidental being has always proven self-defeating, since what is accidental is not related to anything beyond itself. Were there only accidental being there would be only endless dispersion or blind congealing. To expect *all* objects lying around us to assemble the foursome is to tacitly accept a mere plenitude of entities as an ultimate principle. Yet our philosophic intuition ought to tell us that a mere plenitude is not enough to assemble the foursome into an abiding togetherness. I think the problem of necessary and accidental being could be handled by saying that the trivial, accidental objects assemble the *ensconcement* of the foursome rather than the foursome itself. They are dissembled things which seem to lack relevance and purpose because they point up the ensconcing nature of the foursome rather than the latter's abiding togetherness. The items mentioned were "too everyday," and the ensconcement of the foursome is precisely the everyday.

A thing in the Heideggerian sense is not an object over against us which we know and conceptually represent. As the simple unity of the assembled foursome, a thing is "in a certain sense" human and divine, earthly and heavenly. This explains how the "realm of all realms," the Heraclitian illuminative clearing (VA 278), the realm of self-spectacle (*ID* 30), the Dimension (VA 198, 200) can apply to human existence and also to the thing-world.

World and thing are compelled in a mutual configuration as moments of the *Gering*, the assembling boundary of the ringing motion of the foursome. Compelled in the mutual configuration, world and thing do not abide alongside one another; rather they are said to *pervade* one another. In their mutual, pervasive thoroughfare they traverse a middle, a mean in which they are at one and yet each retains its respective integrity. As such, world and thing are contiguous so that an intimate innerness prevails: "As so at one they are intimate. The midpoint of the two is innerness" (USp 24).[17]

This would indicate that innerness and Gering have essentially the same meaning, although innerness emphasizes the intensional aspect, that is, the point of tangency between world and thing. Gering emphasizes the extentional, ecstatic aspect whereby the foursome are assembled into a simple unity, their dynamic character momentarily restrained so that they do not each go their own way and flow into endless dispersion.

Innerness of world and thing is not a fusion; in order for innerness to prevail, world and thing must maintain distinct integrities. Thus Heidegger speaks of a "cut-between," an "inter-cision" (Unter-Schied) (USp 24). The inter-cision gives rise to world and thing rather than being merely a middle ground for two already-established terms. The inter-cision is neither a distinction nor a relation. Cutting-between, it opens up a realm-between such that world and thing are each allowed their own essence and integrity and yet are allowed to pervade one another (USp 25).

The inter-cision is characterized by Heidegger as Di-mension (USp 25). As Dimension the inter-cision measures and weighs out world and thing, each according to its own character. This weighing and measuring out opens up the middle realm—the "Between" (das Zwischen)—in which world and thing are borne apart and yet ordered to one another in mutual thoroughfare. As the mean for these two the inter-cision is the measure of world and thing: the Dimension of their difference but also the Dimension of their self-sameness and mutual, pervasive thoroughfare their innerness.

World and thing are in a sense residual moments of the dynamic inter-cision. Were there no such residual moments the inter-cision would collapse into an amorphous identity. In unfolding the inter-cision brings about world and thing as decisive moments. Because the unfolding/folding is decisive, the inter-cision cannot unfold and then fold back upon itself without "letting the creases show." This accounts for the arrival of things as the residual termini of the ensconced simple fold and thus for historical decisiveness. Things are decisively borne apart from world and world from things, giving us a residual, historical past. This decisive out-bearance (Austrag) occurs at once with the unfolding/folding of the twofold inter-cision. Out-bearance is the sign that the unfolding is not a one-sided revealing;

out-bearance is the mark of the necessary belonging-together of unfolding and folding, of disensconcement and ensconcement.

The out-bearance of world and thing does not merely result in a separation and assembling of these two however. With the arising of world, and therewith of the foursome, which total occurrence is also the out-bearance of world and thing, arises man as he who dwells both alongside the things and as a participant constituent of the foursome. As such the out-bearance is an out-bearance for man, a Difference.

Out-bearance is responsible for the fact that, from the standpoint of man, world and thing appear as quasi-objective. We speak of world in terms of a supervening and of thing in terms of an arriving. Supervention and arrival are defined from the standpoint of the human realm, which is also the point at which the supervention "leaks over" by ensconcingly involuting its simple unity into a colorful multiplicity of things. Only by such an involution can there be a preservation of the simple fold and also an unfolding which culminates in an ecstatic unity rather than in chaotic dispersion.

The unfolding/folding twofold principle may be said to have the character of a reflexive interplay between ensconcing itself qua simple fold, unfolding into an ecstatic, illuminative unity and withdrawing back into the ensconcement of its simple foldedness. This however describes the same motion as the reflexive self-spectacle, where an ecstatic to-show-itself arises out of concealment and involutes back, culminating in a letting-be-seen. What is let be seen is not the pure to-show-itself, but rather the "vestiges" of the involution, the colorful multiplicity of things.

Pure self-spectacle (*Ereignis*), for the moment considered in abstraction from all out-bearance or inter-cision, gives rise to the reflective interplay of the foursome. The four are not independent bodies but have instead more the nature of poles of the interplay, each reflecting and designating the others in itself and itself in the others, much as a series of mirrors engage in mutual self-reflection and self-designation. It should be remembered that we are considering self-spectacle as abstracted away from inter-cision and out-bearance; thus there can be no talk of entities. Under this condition the foursome could be merely empty poles of reflection; there could be nothing to even guarantee their polarity. Without some kind of polarity there can be no dynamic interplay. There is nothing to prevent the foursome from collapsing into either a simple, amorphous mass on the one hand or running away with itself into a chaotic dispersion on the other hand. In either extreme the dynamic self-spectacle would become an empty identity rather than maintaining its own vibrant integrity.

However, if both extremes of the interplay, that is, the simple unity and the ecstatic dispersion, are borne apart from the interplay itself, then the latter is in principle protected from an unequivocal collapse into either extreme. If the two are borne apart and yet held in a mutual configuration, the tension set up by them actually preserves the dynamic character of the interplay. The radical concentration of proximity, held in tension and restraint, takes on the character of a simple, concrete unity: thing. The ecstatic dispersion, held in tension and restraint, takes on the character of a bounded ecstatic unity: world.

Technology and poetry

Heidegger's discussion of thing and world is highly poetic and might seem to have little application to the facts of our world. But it would be a mistake to write him off as seeking a romantic solution that is irrelevant to our present existence. This charge has often been made against Heidegger: that he seeks refuge in the pre-Socratics or in poetry and fails to have a confrontation with the modern world. But Heidegger himself insists that his thoughts on *Das Ding* must be taken together with those on *das Gestell*, which he identifies as the essence of technology. Many of the things Heidegger has to say about technology are harsh, but this does not mean that Heidegger has failed to have a confrontation with the problem.

Heidegger's discussion of the thing belongs to a larger problem which is easily overlooked: as he puts it: "That world show itself as world, that thing become relevant and acquiescent as thing, this is the remote arrival of Being itself" (*TK* 42).[18]

Only when world and thing are brought into the polar configuration (*Gering*) so that each indicates the presence and meaningfulness of the other, can Being arrive in its own truth (as the Difference). Otherwise the dimension of Being (and thus the Difference) is hidden from us and all we see are mere things-which-are lying around us. Thus behind the problem of world and thing looms still the ontological difference.

Our world today is one dominated by technology, yet we must be careful in speaking of technology as though it were some abstract essence, which could be understood in its own terms alone. Any discussion of technology must treat it as an historical appearance, dependent on many factors that make an age what it is. Each age has its *techne*, its ways of bringing-forth and letting-be those things which mirror its character. For the Greeks that *techne* par excellence was art: architecture, sculpture, poetry, drama. For our age it is modern technology.

The primary characteristic of modern technology according to Heidegger is its concern, not with things, but with energy. The world is looked upon as a kind of huge storage battery of energy resources which must be unlocked, put under control, converted to other forms of energy, and so on. The attempt to possess and control energy involves technology in a power struggle. Thus the will behind technology is a desire to master the earth, all things within and on top of it. For Heidegger the fact that technology plays such a large part in political power struggles is no sheer accident. This power struggle is part of the very essence of technology itself.

Modern technology has the effect of doing away with the polarity of world and thing.[19] The technological perspective recognizes only one fact of reality: energy and the possible ways energy can be treated. The "truth" about anything seen from the technological perspective must be measured in terms of the amount of work it can do, that is, its energy potential. Thus the qualitative or distinctive character of natural things is ignored; a towering mountain is only of interest if it contains some kind of useful ore or some other resource of energy, a field is a possible site for a factory or for agricultural industry. Nature, in short, is reduced to "natural resources." The fourfold character of world is entirely subdued from this perspective. Man becomes a stranger to his Region.

Of course some tendency to view things merely in terms of their functions or abilities to produce work has always been with us. Ancient man viewed beasts this way, and sometimes even other men, as when they were enslaved. Yet this tendency to view things this way was never made into a world-view until the present age. We have experienced a technological revolution, an emergence of a world-view which is dominated by technology. Perhaps this is even clearer in the communist nations, especially the Asiatic nations, than in the so-called free nations. Yet Heidegger considers this a total phenomenon;

he considers Russia and America "metaphysically the same," insofar as they both are dominated by the world-view of the technological perspective.

Technology is not an "instrument" of man which he could choose to do without; the technological revolution, like other revolutions, was not something that man willfully entered into but rather something that he found himself thrown into. Heidegger cannot accept the view that "technology is in essence neither good nor bad, because it depends on the use to which man chooses to put it," for this would suggest that the relation between man and technology is an external one. Heidegger not only claims that man and technology are intimately related; he goes on to argue that modern technology is based on a kind of thinking and world-view that is subversive to man's essence. The essence of technology is itself a danger to man.

Heidegger points out the way that technology has overcome spatial and temporal distances and enabled man to control material factors in his environment. But man has not made himself any more at home in the world through technology; instead the technological world has made it possible to become alienated to a degree that was never possible before. Although it is true that distances no longer present real obstacles, we do not have a closeness with the world and the things within it, nor for the most part with other men. Man is deeply puzzled about his orientation; feeling alienated from the things around him, he loses his own sense of identity. Hence the "search for modern man," and yet at the same time the attempt to dissolve him into his technological world.

The driving force behind technology is not man, according to Heidegger; nor is it something technological, as though the technological were a self-determining autonomous activity. Rather, technology is modern man's response to the primordial injunction—a paradoxical response, however, which seeks to deny the very possibility and category of something like a primordial summons or injunction.[20] The question whether this is the "right" response, however, is not a meaningful question to Heidegger, since man does not have the simple choice of arbitrarily creating his own response. How man responds is partly up to him, but his finite freedom is always subject to the nature of the injunction itself, especially in regard to how the injunction discloses itself. Thus man's response is not a simple matter of his own doing.

Heidegger calls the force behind technology *das Ge-Stell*, which summons and assembles the various activities of man having to do with stabilizing, ordering, fixing, determining, harnessing, challenging, and releasing energy. I will translate the term *Ge-Stell* as *positioning matrix*. The positioning matrix is the form in which the ensconcement/illumination presently unfolds itself; as Heidegger puts it: "The abidance of modern technology resides in the positioning matrix. The latter belongs to the fated mission (destinate-historical character) of the dis-ensconcement" (*TK* 25).[21]

Hence the problem could not be solved by "returning to the past." There is no simple return. The only way man can meaningfully confront his technological world is through poetry and reflective thought; these are the only weapons we have against any form of slavery or servitude, which is the working result of a one-track response or reaction to an impetus. Examples of such servitude would be total succumbing to—or even total, thoughtless rejection of—the technological world.

What would it mean to succumb? What is the peril (*Gefahr*) Heidegger speaks of? An obvious peril might be the immanent possibility of destruction through total war. But this is not what Heidegger has primarily in mind. Rather, he is concerned about the way we can and often do indiscriminately allow technology to do our thinking for us. Examples might be finding one's "self" through taking psychological tests, finding a mate by means of a computer match, taking a "test" to see if one has a religious vocation. If we fall slave in this way, man will lose his openness for hidden possibilities; his world will be sterile and devoid of spiritual life. Many modern artists depict such a world in which man becomes lost amidst a towering presence of lifeless, massive structures.

Such a world would involve the entire concealment of the twofold Region, even the side turned toward us. We would be unable to think about the essence of technology, for the limits of the world are shrunk. Our clues to the twofold nature of ensconcement/illumination are based on finding sudden arrivals and departures in our everyday experience. Technology works against these; it turns night into day, controls floods, attempts to bring everything around us into regularity. Hence we lose our clues; although our life can become more comfortable, it can also become a narcosis.

This is why Heidegger says we must, while using our instruments, look beyond them to a possible world remote. We must not fall into the trap of thinking that technology will provide man's answers.

Technology is not only blind without a searching inquiry into its essence and the essence of man, but, more than that, technology is an insidious danger if not fitted into a perspective and held in check.

The hold of technology on our thinking is exemplified most significantly in our attempts to account for technology itself. We say that it is the inevitable consequence of progress. We say it is a creation of man. We say that it is a means to an end. We hope that someday technology will conquer death and disease, that man will create his paradise. Yet these hopes themselves are permeated by mystery and uncertainty. Will man succeed? Will he "use" his technological knowledge for constructive ends? Or will the world be destroyed by the blind power struggle which is the driving force behind our present technology?

The abode or essence of technology for Heidegger is not itself something technological. Rather, it is a manifestation of Being (*Sein*), a mode of disclosure (*Entbergen*). The fate of this essence is part of the destinate-historical character of Being itself (*das Geschick*). This essence is presently closed to human vision, and can only be opened by a stepping-back (*der Schritt zurück*) from the immediate concerns of technology itself and widening our visions. We must hold ourselves open for mystery and ensconcement as well as discovery and illumination. We must acknowledge our kinship to the twofold ensconcing/illuminating Region.

To hold ourselves open for mystery and ensconcement is to participate in the fourfold interplay, so that we are able to dwell in the proximity of these poles while yet remaining remote from them. The meaning of the remote is the arrival or appearance (shining forth) of world. But world can only arrive if we allow things to be in our midst as *things*.

To distinguish between mere objects lying around (*vorliegend, das blosse Seiende*) and true *things*, a measure of thinghood is necessary. This measure cannot be taken from what is lying around, since the measure must be selective. It can only be taken from the very limits of world itself, the foursome. The poet is he who is closest to this measure, and it is therefore he who lets things *be as things*. The narrow sense of Reality as actuality is wiped out, for, as Heidegger puts it, the poet *calls* things and in doing so allows the arrival of the fourfold interplay of world. Heidegger asks: "Which presence is higher, that of things-lying-around us or of things-summoned?" (*USp* 21).[22] The things merely lying around us tend to be opaque and distract us from world rather than allowing us to dwell in the proximity of it. The poet goes beyond what is merely there and calls or summons those things which can come to be in a higher sense, in that they disclose world more transparently and intimately for us.

The technological engineer challenges and consumes, whereas the poet preserves and lets-come-forth. Heidegger contrasts the two in speaking of the Rhine River. He points out that the engineer reduces the Rhine to "source of energy" and builds hydroelectric plants on its banks. The engineer marrs the landscape and makes the Rhine a kind of huge storage battery; he sacrifices the Rhine in its original beauty in order to give us something else. The poet, on the other hand, lets the Rhine be present in a more vibrant sense than it was before. The poet *lets* the Rhine *be*, lets its essence shine forth as not merely a source of energy or a means of transportation, but an assembler and relator of men, a wandering expanse, a giver of nourishment and life, a gift of nature, a separator of man, a destroyer of life. Thus Heidegger in his later works rejects the view that the full significance of things lies in their instrumentality; he insists that we cannot encounter a thing in its essence unless we encounter it poetically, by looking beyond its immediate, actual use. In short, we first encounter the Rhine in its full significance and essence when we see it as a focal point of world: earth, heaven, gods, men. But for this to happen, the Rhine must also be more than an object of visual perception; it must attain the "higher presence" of being called forth, rather than the indifferent presence of something lying around.

But suppose we protest that fiction is wider than fact and that to ignore this distinction would be to lead the life of a maniac. Technology gives us graspable things that make a difference to our everyday lives; mere poetry gives us chimeras and illusions which may be aesthetically inspiring and significant but useless in battling with existence. The argument continues, we cannot make do with poetry alone; we must therefore relegate it to its "proper" place—a kind of aesthetic luxury to be enjoyed when our work is done. This is often put by saying that life includes both the practical and the theoretical. Poetry cannot be "applied" and is thus a kind of luxury.

There are two replies which can be made to this argument. The first is that our work is never done. Therefore it is unrealistic to save poetry for a time which never comes. The second, and more philosophical, reply is that the argument assumes that poetry and technology are through and through dissimilar and unrelated. But Heidegger points out that both have their essence in *aletheia* and therefore belong to the destinate character of the unfolding/folding. Only because this is so could these two conflict. In some respects this conflict is a tragic one. There was a time when poetry and technology (techne) were in harmony with one another: the age of the Greeks and to some extent the age before the modern period, before the dawn of modern science. In the dawn of modern science Heidegger sees present the hidden seeds of the technological revolution and the positioning matrix. With the modern era comes a gradual kind of disparagement of poetry and myth as forms of disclosure, and a dominance of the mathematical, which is applied not only to science but also to art itself in some cases. This was almost unheard of in the days of the early Greeks. Even the Pythagorean numbers are not primarily mathematical quantities but principles of harmony and balance, attunements between man and the void. The mathematics of the modern character has a different flavor, because it is taken to be the art of calculation and schematic positioning.

Yet Heidegger emphatically denies that we can abandon technology in favor of poetry. Such a superficial move would overlook the fact that both are rooted in something more fundamental, namely the twofold character of the primordial injunction. Both belong to the fated mission (das Geschick) of aletheia: ". . . not returning into the origin, but going forward into bare unconcealedness . . ." (2 Ni 458).[23] The positioning matrix or essence of technology Heidegger takes to be the extreme point of the concealment of the simple fold. In other words we are nearing a turning point, as is signaled by the extreme peril in which man's essence is placed. What man must do is confront the peril; he must think out the essence of technology and its hidden kinship with poetry. For above all, man is called upon to think, to penetrate and enter into proximate relation with the essences or abidances of things.

The positioning matrix is not an ultimate; but neither is it an object which man can simply take or leave. As a part of the destinate-historical, the fated mission of disensconcement, it is something which man must overcome through reflective dialogue with that which underlies it.

But man must have a glimpse of his Region, his true abode, if he is going to confront his peril. It is the poet who provides him with this vision of world and thing, who shows him his true Dimension. The poet calls the things forth and lets them be in their highest sense. The man who weaves fantasies and fictions is not necessarily a poet, for the poet is concerned with the arrival of world: the interplay of gods, mortals, heaven, earth. In summoning things forth, the true poet must do this in such a way as to allow what Heidegger calls the "remote arrival of world." By this he means that we feel the proximity of this interplay but do not attempt to grasp it literally as we would a thing. In and through illuminating/ensconcing rapport with things—sometimes grasping them and feeling them in their closeness, but always allowing them to be present as focal points of world itself—we hold ourselves open for the ensconcing arrival of world and mystery. We respect the twofold character of the Region; man's death in fact shows his kinship with the self-ensconcing character of the Region. Thus, man's nature is again shown to be radically bipartite. On the one hand, man is a body, a thing alongside other things. On the other hand, he retains his membership in the foursome; he can ecstatically participate in the illuminating/ensconcing Region (*Gegend*) and hence encounter all things illuminated/ ensconced therein.

World and thing, while remaining poles which are borne apart, manifest a kind of intimacy or innerness (*Innigkeit*). This intimacy involves a dialectic of remoteness and proximity: world arrives in the guise of things and yet ensconces and withdraws its own openness. Heidegger calls this innerness or intimacy *pain* (*Schmerz*) (*USp* 27). Man can never dissolve himself into either empty openness or dumb thinghood; his Dimension remains the locus of the innerness of world and thing. Pain thus has cosmic significance, and this is why it can become a part of tragedy and true poetry. Pain is the rendering of proximity remote and of remoteness proximate; pain opens up a schism (*USp* 235). At the same time pain assembles what is on both sides of the schism. Pain cuts apart and assembles upon itself what is cut; as such pain gives articulation and determinacy to the realm between world and thing. Pain forms and structures the schismatic realm-between, the Dimension (*USp* 28).

The phenomenon of innerness can be taken as a reinterpretation of to-be-in, sometimes called in-stancy (*Inständigkeit*), which had been dealt with in *Sein und Zeit*. However the analysis in *Sein und Zeit* dealt too much with the subjective side of human existence and thus could be easily misinterpreted as a mere ontologization of the subjective consciousness rather than understood as a medium of world and thing. Heidegger tries in his few, cryptic remarks on innerness to define human existence as the *meeting point* of world and thing, rather than starting with our existence as something already given independently of the thing-world. By identifying human existence and innerness Heidegger underscores his contention that the disclosure of ourselves necessarily involves a disclosure of world and thing and vice versa. This, when viewed in retrospect, is easily recognized as the very cornerstone of the analysis of Da-sein as given in Sein und Zeit. This problem, however, I cannot enter into here.

Pain signals a border between our "thing-like" nature and our congruity with world. Pain is associated with death, for death signifies the extreme degree of proximity to this boundary. But death is also the absence of pain, for in death the boundary dissolves. The boundary between world and thing is called a threshold which is held rigid by pain (USp 26–27). The threshold is rigid because we do not cross over to one side or to the other and still remain man. Pain rivets us onto the threshold. Pain signals with ultimacy and finality that we are conditioned, needed and used by a cosmic principle: that we are in the ensconcing/illuminating Region.

Heidegger's approach to the problem of pain can allow us to focus the dialectic between technology and poetry. The technological perspective interprets pain as a practical problem; understanding pain means only understanding its physical causes and attempting to remove or minimize them. Yet the poet and the thinker are both very much aware that many of their fruits are grounded in pain, that without pain human existence would perhaps be a very superficial thing. Most great works of art evoke dialectical interplays of pleasure and pain, joy and sorrow. Yet only perversity is won by deliberately cultivating pain; pain must not be blindly sought nor avoided but rather understood. Rather than taken as a principle of explanation, as Hedonists and Utilitarians do, pain must itself be explained. What is its ontological significance? Unless we can answer this question in some fashion, we can hardly expect to determine what man's proper attitude toward pain should be. And we certainly cannot take for granted that pain should be regarded as simply a challenge for technology.

Pain for Heidegger, is not something merely clinical; rather, pain is a form of disclosure. Only a being that is "between," in the self-ensconcing Region and yet also alongside the things, can feel pain in its full sense. Pain discloses to us this duality; we are alongside things in that we are vulnerable to them, but we are "not of their world," since the pain may linger even though the original disturbance has long vanished. Some of the greatest pains are not caused by intrusions of things at all, but by withdrawals and absences. In experiencing a withdrawal, we are drawn towards the empty openness of the ensconcing side of the Region, but we are not permitted to enter

The Realm of all Realms 155

its simple interiority; pain rivets us on the threshold, where remoteness and proximity meet and are borne asunder.

Let me now attempt to reword this in more conventional terms. Pain tends to constrict our vision, to immobilize us; pain can haul the most noble thinker out of the ecstatic realm and make him double over. Thus pain discloses the way in which any ecstatic traversal of the open realm must also involve a return to one's facticity. Yet pain can also destroy our absorption in any particular, ontic set of circumstances and thus act like the call of conscience. If we are absorbed in one of the poles of proximity (things around us) or remoteness, (world, the open realm) pain discloses the relevance of the other pole. Now on the one hand man could try to anesthetize his pain, to avoid it by dulling himself to it. But to do so would be for Heidegger to succumb blindly to it and to forget the fact that pain signals a border or threshold between world and thing. If man were not congruous with this border, he would feel no pain. Plants and stones feel no pain. Animals feel only physical pain and sentiment anxieties; but man has spiritual pain, pain produced not by this or that thing but by the hazardous openness of Being itself, and this is in some form always with him. To completely subdue pain, he would have to reduce himself to a stone. Yet on the other hand, if man made no attempt whatsoever to deal with pain, he would be equally incapacitated. It would be foolish for a thinker with a toothache to refuse to go to the dentist and to rather sit around glorifying the nature of pain. Heidegger is surely not advocating this. But we must attempt to understand the ontological nature of pain and learn when and where to avoid it and when we must face up to it. Some kinds of pain may be more significant than others. The pain the mind feels in trying to grasp infinity may be more important than the pain we feel with a toothache. Should both be looked at as merely clinical or pathological? Is the view that philosophy itself is pathological an extreme symptom of the failure to confront pain as an ontological phenomenon?

VI

Language and the Ontological Difference

... nun aber nennt er sein Liebstes,
Nun, nun müssen dafür Worte, wie Blumen, entstehen.
(*Hölderlin*)

The Reduction, the step backward, whereby we gain access to the Region (*Gegend*) or heart of the unfolding/folding involves an overcoming of the metaphysical perspective and its bifurcations of the "actual" and "possible," the "real" and the "unreal," the "true" and the "false." This involves developing a more authentic outlook which recognizes the relevance of the twofold unfolding/folding and the mysterious interplay of its poles or moments, world and thing. But how do we apprehend such matters?

The Region cannot be seen, since it is self-ensconcing. What is seen must appear and thus be disensconced. We can only see the side of the Region which is turned toward us, the transcendental horizon in which all things-which-are appear. The simple foldedness or self-ensconcing character of the Region eludes our vision. Then how do we know that there is such a withdrawal?

The possibility of such an apprehension only opens up when we give up the primacy of sight as a means of acknowledging the presence of things. How do we know that elusive things are elusive? Heidegger's answer is that they call to us. A call can have a mystique about it; we speak of the call of the wild, the call of adventure. Here one knows he is being called, but the source and character of the call remain undetermined. Even a call for help can draw us on into regions of uncertainty and darkness. Heidegger portrayed the ontological difference as a kind of injunction or call to which man must respond. Man's response takes the form of thinking: taking note of, gathering and preserving things-present in their presence. Yet thinking is not some kind of ghostly occurrence inside heads; thinking is a dynamic traversal of the open realm; it necessarily involves the appropriation of language, whereby things-present are gathered and acknowledged to be present.

For Heidegger there is, as has been seen, a higher and a lower sense of "presence." Some things we merely let lie around before us; we are indifferent to their presence in that we do not view them as reflecting the interplay of the world-foursome: earth, heaven, mortals, gods. The highest kind of presence is deliberately evoked or called forth; in this case a thing is present as thing, that is, as focus of the interplay of the world-foursome. This higher kind of presence is in the charge of the poets and thinkers, for they are the ones who are attuned to the four world-poles. The everyday or calculative understanding falls short of recognizing earth, heaven, mortals and gods in their fullness. Yet even there the role of language as *logos*, letting-come-forth, is unmistakable.

The unfolding/folding of the twofold is the occurrence of the illuminative clearing, and this has already been linked with self-spectacular appropriation (*das Ereignis*). Thus Heidegger says: "The prolific appropriation, which excites or moves Saga as the token in its showing, can be called self-spectacle. It gives forth the free openness of the illuminative clearing, in which things-present linger in their duration; out of which things-absent escape and in withdrawal still keep their continuance" (*USp* 258).[1] We read further: "Saga, which resides acquiescently in self-spectacle, is as to-show the most apropos mode of self-spectacle. Self-spectacle is locutionary in the sense of oracular" (*USp* 262–63).[2]

Language as to-show, to-let-appear, lets world supervene, wherein the things arrive. For Heidegger there is no such thing as first an occurrence and then a naming apprehension of this occurrence. The occurrence of red and the naming apprehension of red are both rooted in a more primary occurrence which is neither "subjective" nor "objective" in character. It is occurrence par excellence, *Ereignis*, radically singular in nature (*ID* 29).

Just as self-spectacle must culminate in an out-bearance of world and thing, so too language as *Sage* is brought to the spoken word. *Sage* itself abides as the primordial summons; through mortal speech the primordial summons is heeded and complied with—whether we realize this or not. Mortal speech is thus characterized as a co-respondence (*ent-sprechen*) or a response (*Nachsagen*) (*USp* 32, 266).

The Region is the realm of the word (Gel 49). Elusive and forboding, ensconcingly illuminative, it calls to us. Yet the call itself is a silent one. It cannot be put into words; only the human responses to it can. To understand why and how Heidegger thinks this is so is the task of this chapter. Since much of this theme has already been implicitly traced, there will be a certain amount of backtracking in order to bring into focus Heidegger's treatment of language. However, this will not be mere repetition, because the ideas discussed will be treated from a slightly different perspective.

Heidegger's philosophy of language

Heidegger's writings on language, collected under the title *Unterwegs zur Sprache*, are among the last works he has had published. Up until that time there are references to language and its problems scattered throughout all his works. His interpretations of the pre-Socratics deal with the ontological character of the Greek language as a central theme. Heidegger himself contends that the problem of language has always been a major focal point for all his thinking; from the very beginning of his thought he has stressed the kinship between language and being (SZ 27–38, 160–67; USp 127; Gel 49). Yet even in *Unterwegs zur Sprache* it is difficult to say whether Heidegger has a "philosophy of language" in any complete sense. His work is sketchy and focused mainly on one problem, the nature and cosmic origin of language. Yet concerning this problem and its implications his thought is searching and philosophical in orientation. For this reason I will speak of his having a philosophy of language and in this section attempt to formulate it.

A language (Sprache) in Heidegger's sense of the term must be spoken, and it must belong to a historical people. Thus German, English, and Gothic are languages, but Morse Code, Esperanto, and symbolic logic are not. Language for Heidegger is factical rather than artificial, and a people's involvement with its language is a total one. Language is not a mere means of communication; a language must express or reflect the way a given people or culture exists. A given language reflects all modes of expression which pertain to the important phases of that people and culture: religious, familiar, poetic, formal, commercial, uncouth, etc.

Language and Ontological Difference 159

For most philosophers the crucial problems in philosophy of language lie in the area of semantics—the relation between language and the nonlinguistic features of the world which are somehow made accessible for us through language. Heidegger would not deny that this is a most important area in philosophy of language; but for him there is an even more important problem, the essence or ground of language itself. He notes a kind of impatience in many thinkers with this problem; indeed most do not ask such a question but simply take it for granted that language is essentially some kind of system of symbols for the purpose of communication. Already in *Sein und Zeit* Heidegger showed dissatisfaction with this kind of treatment of language: "In the end, some day the philosophical undertaking must take it upon itself to ask, what sort of being applies to language. Is language an instrument within the world, ready-at-hand, or does language have the kind of being which pertains to *Dasein* or does neither of these two alternatives apply?" (*SZ* 166).[3]

Why is this question so important? If language is a tool, an instrument, this might suggest certain important analogies with tools, such as: that their primary value is their usefulness, that is, they have no value in themselves and could thus become obsolete. The horse carriage was an instrument of conveyance—a vehicle. Moreover, this vehicle conveyed other (now almost obsolete) instruments—for example letters to friends. Today, although the need for vehicles is not obsolete, the horse and carriage is obsolete. Because the horse and carriage is obsolete, however, the functions of the blacksmith and liveryman are obsolete. This example serves to raise the following questions, by analogy, about language: (1) Can we say that language "has" a function in the way we can say this about vehicles or blacksmiths? (2) Granting the possibility that words "have" functions and that words can become obsolete, would language per se become obsolete? My suggestion is that there may be some basis for an analogy between *words* and tools, but would that relation necessarily hold between *language* and tools? (3) Can man transcend language, gain mastery over it and shape it to his arbitrary needs in the way he apparently can his tools?

On the other hand, if language is part of being human (*Dasein*), then we would expect certain analogies here too. For example, being human is historical, aesthetic, religious. It is possible for men to *die*; but dying, I should hope, is rather different from becoming obsolete. Words can become obsolete; but languages can die (*SZ* 166). We would expect that, on this view, language would be bound up with the very destiny of man, not so much mastered as appropriated and attended to.

In Heidegger's later work he changes the emphasis of his question a little. The issue is not one of tool-being versus human-being, but rather one of the relation between man and language. In *Sein und Zeit* it sounds sometimes as though a kind of ontological classification is going on; but the later Heidegger is not content to divide what-is up into kinds of being (*Zuhandenheit, Vorhandenheit, Dasein*) in quite the way he was earlier. The issue later becomes for him not one of kinds of beings but rather *how* we relate to these things and *how* they are present for us, be they cups, poems, or persons (VA 145, 163). I think this is what was also intended in *Sein und Zeit*, although the problem is perhaps in some ways misleadingly stated.

It is Heidegger's view that the functions of language pervade all moments of human existence. Language is used to communicate information. It is a medium of emotive expression. Language is also the medium of private conversations with ourselves or wrestlings with our consciences. Our memories release words and voices of the past. We give ourselves away, we conceal and deceive through language. Language discloses the multifarious aspects of our existence—sometimes even when we don't want it to. Man's relation to language cannot be exhausted simply by saying man *uses* language (whatever that means!); for sometimes it would seem that language uses us, as when we find ourselves saying things we didn't mean to say and when we make puns.

Language allows us a certain mastery over things and over ourselves. Without words our world would be a darkness of immediate chaos (USp 177). At best our attention would always be riveted on the immediate; for things remote must be *called* to our attention. Moreover, without language our only relation to what is and what goes on must be one of blind passivity; for without language we cannot stabilize anything or impose controls or relations. Moral philosophers have long seen the connection between deliberate actions and imperatives which underlie them and make possible their intelligibility. Without language we could not choose to call to mind certain memories or sets of experiences rather than others; thus we could not deliberate any future action but would be victims of vague hauntings and would stand by helplessly witnessing what we did, no matter what it was.

But this does not necessarily exhaust the nature of language. In particular, it does not yet entitle us to the "pragmatic," utilitarian view of language, according to Heidegger. It is interesting to note that part of Bergson's attack on language is due to his instrumental, utilitarian view of it. Heidegger, on the other hand, makes the claim that this interpretation of language falls short of the full essence of language. Hence the moving power of language is weakened or lost, and we have the appearance of static words and meanings and a grand *ars combinatoria*. The poet especially cannot treat language as a mere instrument; one of the features of good poetry is that it brings the richness and beauty of language to light. For the poet language is his element as well as his tool; for this reason poets and poetry are given a great deal of attention by Heidegger. He claims poetry provides pure examples of language because in poetry language operates as an unfolding power rather than as a mere instrument. This claim is, of course, in need of examination, but first it is important to see clearly what Heidegger's claim is and what implications can be drawn from it.

Heidegger rejects the "pragmatic" account of language for several reasons. To begin with, the relation between instruments and men is in much need of clarification. It is clear that a hammer is an artifact, external to my being; my interest in the hammer is basically a practical one. (Heidegger has much more to say about this, but I cannot enter into that here.)[4] But what about my hands? They are more than just instruments; they are part of me. My hands are not merely useful; they are expressive of my moods; sometimes I adorn them as objects of vanity; my hands are much closer, more natural to me than are the instruments of instrumental theories; yet at the same time it is much more difficult to characterize my relation to them (*WhD* 51). The same holds, for Heidegger, in the case of language; being at home in a language is part of being human. But there is an even more serious objection to the instrumental view of language. If language is an instrument for conveying meanings, we must at some point suggest that there are some meanings intuitively graspable without the need of language—otherwise how would we know what we wanted to say before we had said it? On this analysis, it would seem that at some point language presupposes meanings; but then the problem becomes one of the relation between language and these prior meanings. Now if these meanings are nonlinguistic and graspable in their own right, language becomes merely an instrument for gossip, since we would be implying that nondiscursive or intuitive thought could occur at the highest and most fundamental level. Language would then be unnecessary. But most philosophers today would be unhappy with this view and would rather hold that human thought is through and through discursive and can only exist as articulated in some symbolic medium. The need for a symbolic medium is inherent in all positions which emphasize human finitude. Also, the attempt to question the nature and status of these "non-linguistic meanings" leads into difficulties such as the third man.

Another alternative is to say that language itself is twofold; that an ontological language underlies our various, ontic languages. Plato and Leibniz would be examples of this position. The important problems would then lie with the ontological language which makes possible the sayable. My ability to say something like, "The door is closed," or to translate that into another language presupposes an understanding of ideas above and beyond the words I have actually used to mean what I say. These ideas are however approximated by my words, and thus when I speak I am always operating on two levels. The problem for philosophy of language, on this view, would be to bring to expression this ontological language and determine our relation to it. Heidegger's position is perhaps most closely related to this one, although his paradigm for the ontological dimension of language is poetic, whereas Plato and Leibniz used mathematical or logical paradigms.

When Heidegger asks about the essence (*Wesen*) of language he is asking about the original dimension of meaning which yields forth particular languages. He is pursuing the problem of the dual level of disclosure: *hermeneuein* and *apophansis* and the spill-over from one level to another. He rejects the idea that we can give a philosophically adequate treatment of language by formulating its rules and formulas. There is the obvious difficulty that no two languages follow exactly the same set of rules. But the real objection is that this approach must use a set of linguistic rules in order to make its formulations concerning the language under investigation. We are thus in a circle: we must use (and thus presuppose) language to inquire about language. An adequate approach must take this fact openly into account, since this is what singles out the problem of language. To go "outside" the field of language would be to lose language altogether.

Openly within the horizon of language, Heidegger attempts to explore the location and the bounds of this horizon. How is language present? Is language an instrument of man? Or is man an instrument of language? How can we best characterize the relations between language and man, between language and the things disclosed by language? What is the connection between language and national destiny, that is, French, Indian? What does language really disclose?

Man and language

In the essay entitled "Der Weg zur Sprache," Heidegger analyzes the anthropological view of language. He claims that this view assumes that language is an "activity" of man, as though this somehow explained its nature or as though man could somehow be what he is outside of language. Any humanistic view moreover hearkens back to the instrumental view of language.

Heidegger attempted to formulate his own view of the relation between man and language in his "Brief über den Humanismus" through the sentences: "Language is the house of Being. In its housing dwells man" (Hum 5).[5] This statement is troublesome for, aside from the fact that houses appear to have very little in common with language, it appears as though there is something smacking of the instrumental view which Heidegger claims to reject. Heidegger himself later called his analogy "clumsy" (unbeholfen) (USp 90). Houses themselves, however, are not mere instruments according to Heidegger, if we appreciate their fullest character. A house, in the fullest sense, is much more than a material building in which to keep belongings, sleep, take meals, and procreate. True houses have histories and distinctive characters. A genuine house calls together the major things associated with human life: security, comfort, family, friends, rest, meals, leisure, as well as the difficulties and problems connected with all of these. But it is through language that man expresses and articulates his concern with all of these. Man most truly dwells in language. Language is man's dimension, his element. Dwelling is not a "function" of man but the way he is, upon the earth, under the heavens, and so on (VA 149-52). And dwelling itself is contemplative through and through, insofar as any genuine dwelling takes place. Dwelling cannot be accomplished mechanically.

Were language not ontological, it could not disclose anything pertaining to the world; there would be no objective grasp of things, no public, overt world (USp 177). Language cannot be just another thing (have ontic, that is, psychological status), for then it would not be a discloser but itself in need of being disclosed. But of course language must have some ontic features about it or it would not be speakable or writable or audible. It would be impossible to make verbal utterances, were language not in some sense ontic; but it would be impossible for these utterances to disclose anything if language were not also ontological. This difficult juncture between the ontic and the ontological Heidegger sought to express by the phrase *Haus des Seins*.

Language, then, for Heidegger, is twofold, and man's relation to it is one of dwelling. To dwell means to abide and to give attention and expression to our abidance. Authors know how to characterize ways of life in terms of one's abidance and abode. People express what they are through the things in their abode and their relation to them. An abode or house provides the open realm or locus in which to articulate one's way of life—what and how one is. The most fundamental abode, the one man can never escape and still be man, is language. Language provides the basis and medium for meaningful contact with everything else.

Man dwells in language through talking and also through hearing. It is, according to Heidegger, the phenomenon of hearing that connects talking and understanding. Hearing is what a child must learn before he can speak rather than just babble. By hearing Heidegger means much more than the psychological mechanisms therein involved; he is referring to a purposive, self-aware mode of behavior. Hearing, as Heidegger understands it, does not give us mere sounds but meaningful sounds—the wail of a child, the sonorous drone of the saxophone, the rustle of leaves, the mysterious hum which *has* an origin which no one can presently locate. Heidegger tells us: "Hearing constitutes in fact the primary and authentic openness of *Dasein* for its own ability to be, as hearing the voice of the friend that each *Dasein* has constantly with it. *Dasein* hears because it understands. As understanding and being-in-the-world with others, *Dasein* is in bondage, both with regard to being-with-others and itself, and thus belongs (*ist zugehörig*) to this bondage or order of hearkening" (SZ 163).[6]

If something is said, it must be taken up or preserved or else it gets lost. The possibility of meaningful disclosure is thus predicated on hearing what is said as well as saying something. At the very least, it must be heard by the sayer. There are well-known sayings about those who talk to hear their heads rattle, who don't even listen to themselves—people who talk a lot but don't say anything, who chatter endlessly but without understanding. And those who lack understanding are usually those who are unwilling to listen, who are easily bored; they drown out the voices of understanding with their own senseless chatter.

To listen, one must be able to stop talking, to exercise restraint. To talk thoughtfully we must listen to ourselves. Often this causes us to falter (zögern) (USp 119). The reason for this, according to Heidegger, is not merely psychological; inherent in the nature of language itself is a withdrawal, a silence. In later works, Heidegger refers to this as die Stille (USp 29–33, 216). From Heidegger's point of view, we could say that Wittgenstein, for example, experienced this withdrawal when he realized toward the end of the Tractatus that philosophers have generated their characteristic problems by trying to say the unsayable—by trying to talk about absolutes and principles and essences as though they were discrete things. But, from Heidegger's point of view, we would also have to say that Wittgenstein did not only go silent, but he let his silence degenerate into a dumb speechlessness. Wittgenstein treated the silent, the withdrawal or refusal, as falling *outside* of meaningful language. He turned away from it. This is typical of positivism which sees breakdowns as occasions for repair rather than as situations of disclosure.[7]

To put this more simply, Heidegger would agree with those who say that everything is expressible and that paradoxes result when we try to say the unsayable. Silence, for example, is unsayable. However, the line between the sayable and the unsayable is not always sharp. Through the use of elliptical and figurative language we can border the unsayable, point to it. Language itself—as language—is an inexpressible; we must "use" language to put things into words, but we cannot put language into words. Language per se transcends any single instance of itself. Language itself remains hidden, silent, as what is said gets said.

What gives Heidegger the right to speak of these hiddens and silences? How do we know there is anything more to language than the actual words and rules of usage with which the positivists concern themselves? Heidegger bases his claim on two points. First, as we have seen, he argues for the need of an ontological dimension of language, which is operative but never directly manifest in the way that things are manifest. Second, as we have also seen, Heidegger devoted a great deal of Sein und Zeit to exploring dimensions of human existence in which withdrawals and silences play an important role. These center around the phenomena of anxiety, death, and conscience. The analysis of conscience is, I think, one of the most brilliant sections of Sein und Zeit as well as one of the most important for Heidegger's later views on language. The important thought to note here is that man is called to authenticity by a call that has no "content." It is a call and yet it is silent.[8] Heidegger does not intend to speak metaphorically when he speaks of the call. He insists that this phenomenon is not merely psychological but ontological; it belongs to the very way man exists. The call is to him "real" because it has real effects and therefore cannot be dismissed as a metaphor. Implicit then in Sein und Zeit is the view that man dwells in language, since he becomes "authentic" by responding to the silent call, and he becomes preoccupied with the particulars of the thing-and-fact world when he does not hearken to the call. This call is a kind of unifier of human thought, it is because men acknowledge the call that their language can have objective meaning and reference. The call then is roughly analogous to the Heracleitian Logos. And the true dweller is the true listener, who hears and then responds with thoughtful and attentive speech, the expression of which is always a function of his factico-historical situation.

Language and essence

Some readers of Heidegger feel that he is most dogmatic and unrigorous when it comes to his treatment of language. They accuse him of laying things down by fiat, of uncritically accepting what certain poets have written about their experiences with language. In opposition to this I will argue that not only is Heidegger's treatment of language careful and rigorous, but it exhibits his stress on the inner unity of phainesthai and logos at its best. In the essay entitled "Dialogue between a Japanese and an Inquirer" Heidegger himself refers the Japanese back to his chapter on phenomenology in Sein und Zeit (USp 95).

We cannot gain a proper understanding of the unfolding power of language nor of our relation to it simply by talking about the formal structures of our language. Language cannot merely be talked about but must be encountered as coming-forth, as phainesthai. Yet language is not a phenomenon in the sense of something-which-appears; a deliberate focus of attention of some sort is required in order to encounter language in our proper and full relation to it. Here Heidegger also points up the importance of logos, letting-come-forth-and-be-present and hence the inner unity of *phainesthai* and *logos*. As we have seen, in order to let-[ ]-come-forth, we must gain access to the proper region in which something unfolds. This leads Heidegger to ask, what is the proper Region of disclosure for language? If we are not at the scene of this disclosure, our attempts to think through the nature of language will be fruitless.

The inner unity of *phainesthai* and *logos* was also seen to lead Heidegger to stress the hermeneutic nature of *logos*. Before something can be attentively thought about, the realm within which it is to be found must be predisclosed in such a way as to guide and structure our inquiry. This means that the disclosure in question is not perceptual but rather discursive in character: a kind of message or annunciation of a field of meanings and structures within which to couch our interpretation of the item in question. *Hermeneuein* thus points up the open realm, within which all things are encountered and suggests that the openness of this open realm is most intimately connected with *logos* itself.

In his essay on Hegel's concept of experience (*Erfahrung*) Heidegger also stresses this idea of a predisclosed intelligibility (the Absolute as *Parusie*) which is necessary at the outset before any kind of thematic investigation can take place (*Holz* 120, 170–77). Because of the ontological difference there is a hiatus between *what*-comes-forth and its coming-forth, between what is in the open realm and the openness of the realm. This openness or Region must announce itself in its own characteristic way.

In the same essay Heidegger points out how the philosopher must not rest content with the results of ordinary seeing. Sight is our ontic sense *par excellence*; above all, it is our sense of sight which acknowledges ontic objects. The philosopher thinker must transform his seeing into a *skepsis*, a seeing-to, a seeing-after (*Zusehen, Nachsehen*) which focuses not on the thing-present but on the character and status of its presence (*Holz* 139–40, 173). Ordinary sight and seeing can even be barriers to philosophical thought, because they rivet us on the ontic objects but are insensitive to the openness of the open realm itself. Ordinary sight can only reveal what is already predisclosed; it cannot originate an original disclosure of the openness of the open realm as such, of the Region. In this openness there is "nothing to see."

Yet the openness or Region draws us on and into it; it has the character of a mysterious call, an injunction. More and more Heidegger moves towards an inner unity of the Region or self-ensconcing open realm and the underlying nature of language. But this move is a direct consequence of the inner unity of *phainesthai* and *logos*. In his dialogue between an inquirer and a Japanese he explains *hermeneuein* as: ". . . that exposition, which brings news, in so far as it is able to take heed of a message" (*USp* 121).[9]

Heidegger also is fond of using the phrase *Geschick des Seins*, the destinate-historical or fated mission of being. Here he underlines the way in which Being itself must be hermeneutic in the sense of oracular if we are to have any important relationship to it. He tells us: "The destinate-historical character or fated mission of Being is as exhortation and demand that decree (*Spruch*) (or injunction) out of which all human speech speaks. Decree (injunction) in Latin means *fatum*. But *fatum* as here construed, namely as the decree or injunction of Being in the sense of the self-withdrawing destinate-historical or fated mission is not something fatalistic . . ." (*SvGr* 158).[10] Always, the emphasis is on the need for a realm of predisclosed meaning—a primordial injunction or oracle, a Region of meaningful presence—in order that any singular effort in thought should be fruitful. We cannot approach something without adapting ourselves and our inquiry to the nature and character of that *within* which our item of investigation comes to be present. Above all, this holds for the case of language. This is the very essence of the phenomenological method as Heidegger saw it from the beginning of *Sein und Zeit*.

In the case of language he applies the same kind of thinking: "If we consult language about its essence, then language itself must already be addressed to or spoken-to us. If we inquire about the essence, namely, of language, so must also the meaning of essence (*Wesen*) be already spoken-to us" (*USp* 175).[11] Were this not so, we would never be able to understand the least thing about language. And if the character of language were entirely opaque to us, then language could hardly be a discloser and communicator of things and their presence. It would be reduced to a set of totally mechanical sounds like the scraping noise of a pipe or the crashing sound when a glass is dropped.

The problem now is to discover the Region in which language comes forth in its most primordial and undissembled character. First we must understand more closely the relation between the Region and our thinking. Heidegger contrasts method, as used in the sciences, with Region (*USp* 178–79). A method is a way of access, but in the sciences the method is formal and thus can be indifferently applied to a variety of subject matters. The method does not create its own content; it is indifferent to its subject matter. For this reason there is a kind of unselectivity to scientific inquiry; it tends to become research into this or that, and there is often all too little reflection on whether this or that is worth researching (Holz 76–78; USp 178). The Region, however, both provides ways or paths for thought (be-wëgen) [12] and provides the undissembled content necessary if thinking is going to be a fruitful endeavor. In other words, for Heidegger it is not enough to be in search of a method; the best method in the world would give us sterile results if we are not at the proper scene of disclosure.

What and where is the proper Region of language? When and where we speak? In *Sein und Zeit* Heidegger had pointed out the way in which the things we use are taken for granted and not allowed to be present except as acquiescent in a context of use. In a previous chapter it has been indicated how Heidegger regards it necessary to overcome our instrumental relationship to things if we are to let them abide in their full significance. The same applies for language.[13] "But, whenever and however we speak a language, the language itself is never brought to word" (USp 161).[14] The relation between language and things-brought-to-word is the same (*das Selbe*) as that between the openness of the open realm and what is in the open realm, between presence and things-present. At the very heart of the problem of language lies the ontological difference. There is a withdrawal in language, and it is at the scene of this withdrawal that language can be most authentically encountered. "But where is language itself brought to word? Strangely enough, there, where we cannot find the right word for something which concerns us, rips us toward it, afflicts us or ignites us with enthusiasm. Thus we let that which we mean or intend acquiesce or rest in the unspoken and thereby, without properly reflecting about it, pass through moments in which language itself has touched us, fleetingly and from afar, with its essence" (USp 161).[15] Being at a loss for words is not a mere psychological phenomenon to be dismissed on the ontic level. Such a "loss" could only occur, given the relation to language that we apparently have. Language is more than an instrument; it has a life and power of its own, and if we are insensitive to this, our attempts to speak will be misfortunate.

On the one hand language has already been addressed to or spoken-to us (*Zuspruch*). Yet this *Zuspruch* carries within it a withdrawal. How is this possible? The full answer cannot be sought anywhere but in the nature of language itself, in the Region where language is at its purest. This Region, Heidegger tells us, lies in the neighborhood of thinking and poetry. This point of proximity Heidegger calls Sage, which means literally Saga. He tells us: ". . . sagan means to point or show: to let appear, illuminating-concealingly to set free so as to proffer that which we call world. The illuminating-disguising, veiling extension of world is that which essentially abides in Saga [Sagen]" (USp 200).[16] It has already been seen to what extent Heidegger views the poet as he who lets things be present in the higher sense, namely as things called forth which reflect the world-foursome and the four world-poles of the interplay. The fact that the poet calls things forth suggests that he has a special relation to language. How is it possible for the poet to call things forth? Because he is more "adept" at language? What is meant here by "adept"?

It is precisely the poet who does not treat language as a mere instrument; part of his "art" is to allow the beauty and movement (Bewegtheit) of language to unfold. He is sensitive to the subtle undertones of language, those tones which the everyday ear fails to hear. Above all the poet is a listener; were this not so, he could not be sensitive to the lyrical quality of a language (USp 70). A good poet then has a closer relation to language than does someone displaying the everyday instrumental outlook; the poet looks at language as not only his instrument but also his element (Hö 33). The poet treats language as the Haus des Seins. Yet not all poetry is of equal quality; only thoughtful poetry reflects the requisite disclosure of language. Such thoughtful poetry or poetic thought occurs only at the point of proximity between poetry and language: Saga, Sage.

By a point of proximity Heidegger does not mean to suggest that thought merges into poetry or poetry into thought anymore than Aristotle meant in the case of composite ousia that matter merges into form and vice versa so that the two become indistinguishable. In the "neighborhood" both poetry and thinking must retain their identity. He fastens on the idea of neighborliness. Two things could be close by one another and yet be mutually strange. What is needed for true neighborliness is a vis-à-vis (gegeneinanderüber). Thus Heidegger says: "The essential presence or unfolding essence of proximity does not reside in the nature of distance but rather the path-making and movement of the vis-à-vis-ness of the regions of the world-foursome" (USp 211).[17] Language then is not simply something in the hands of man, but rather the unfolding, path-making occurrence of the interplay of the world-foursome. Man participates in this interplay by responding to it; the interplay operates or acknowledges itself to him as the primordial call, albeit the call of Silence itself. "We call that soundless silence calling gatherage, that as which Saga moves and makes paths in the world-relation, the resound of stillness. It is: the language of essence" (*USp* 215).[18] The language of essence, the essence of language belong to the same Region. Both are predicated on the silent call of the Region itself, the silent interplay which man is called to play along with. Sometimes we are fortunate in our playing-along, and our efforts yield gems of insight and thought. Othertimes the results are barren. Yet we are at the mercy of the play, for "The play is without a 'why.' It plays, while it plays. There remains only play: the highest and the deepest" (*SvGr* 188).[19] The rules of this interplay cannot be formulated; for what would we use to formulate them? Thus, although language can be useful for formulating rules about things or about certain aspects of communication, the primordial essence of language as rule-giver remains without rules. It is "answerable to itself alone" (*Gel* 49). And man is answerable to it, and answers through his speaking response to the primordial call therewith being drawn into and through the open realm.

Word and thing

Language, according to Heidegger, underlies and makes possible any encounter of things-which-are; as such, language belongs to the Region, that which opens up a field in which things can be encountered as present and meaningful. But in order to do this, language must somehow carry over into the ontic; otherwise it remains an impotent abstraction. Something of language must occur on the ontic scene. This, as we have seen, is accomplished through mortal speech. Mortal speech must articulate words. Words, then, apparently have to do with the finite aspect of language.

But at this point two questions arise: What is the relation between words and the primordial Saga? What is the relation between words and things? The first question cannot be answered independently of the second. Yet a few remarks are in order now, I think, in order to set the tone for the discussion of both questions. These are primarily of a negative nature.

It won't do to consider language an abstract totality of words. This would imply that language is a static abstraction; I can think of no reputable philosopher who would hold this simple view. Words must be grounded in something which provides their meaning, for example, rules of usage. Thus our question becomes: How are rules and words related? Are rules of usage themselves words? Must rules of usage be formulated? Can all the rules of usage be formulated without involving an infinite regress or circle? What would such an infinite regress or circle mean?

It would seem that language is not *given* as words are. Words we find in dictionaries, but we do not find languages in dictionaries. Yet are the words in the dictionaries the same words we speak? If so, then words are not material, physical things; for the latter can be only one place at one time. But we do not literally take words out of dictionaries and then speak them. Where are the words we speak when we aren't speaking them? What are words?

We tend to think of words as purely external manifestations, primarily audible or visible, which have no interiority of their own. To put it another way, we look at words as having only a purely representative character, so that words themselves do not *disclose* any information, but meaningful combinations of words may be "about" something and thus *convey* information. Heidegger, however, attributes to words a kind of interiority; they have histories, for example. Words are fate-laden (*geschickt*); we are not free to use words just any way we please, unless we wish to babble nonsense or deliberately amuse and confuse everyone (*Holz* 300–3, 310–34; *WhD* 84). A word suggests other words; it thus includes within it an entire family of meanings. Here Heidegger does not have in mind something so arbitrary as "word association." Rather he has in mind fields of meaning which can be objective and historically encountered, as for example, when we look up a word in Grimm's etymological dictionary and see what uses it has been put to in the past or when we look up a word in Roget's Thesaurus and discover what other fields of meaning overlap with it.

Words do not stand for things or invariants, and they cannot be fitted out to concepts. Heidegger tells us: ". . . words are indicators and not signs in the sense of mere labels" (*USp* 119).[20] This means that language cannot hope for logical precision, but it also means that words themselves can disclose hypothetical information, in the sense of hints or indications about reality. Through words our human dimension becomes structured, but we must always be careful not to presume that we can completely put Reality into words.

Words disclose but they also conceal; they are hints, not literal descriptions.

According to Heidegger, language does not fulfill, primarily and originally, a descriptive function. To be able to describe something presupposes considerable familiarity with things *like* the thing being described. We say we must be familiar with the concepts being used in order to understand the description. This is where concepts have their place—in that aspect of the world that is generally well-defined so that our knowledge is primarily a re-relating of things we are already familiar with, in other words, the apophantic level of disclosure. Ordinary language and explanations, as well as most scientific ones, presuppose such a world which is already prestructured and pre-understood. How is this possible? What ultimately lies back of this seeming self-evidence of our descriptive language? How is it possible for our language to be *about* anything?

Heidegger's suggestion is that the rules of thoughtful and attentive usage of language belong in some sort of harmony or sameness with the laws that govern the presence of all things. This harmony or self-sameness Heidegger calls das *Ereignis* (*ID* 13–34). This bears a close resemblance to Kant's transcendental deduction of the Categories, where he maintains that "the conditions a priori of any possible experience in general are at the same time conditions of the possibility of any objects of our experience" (*Kritik der reinen Vernunft*, A 158, B 197). What it means is that, at any and every moment, we operate in a world whose general character is predisclosed to us as, we might say, a realm of relevant possibilities. It is only because language on the ontological level has already made this world of meanings and now familiar objects accessible to us *as* this or that object that our parlance within this field can be said to be properly descriptive or not descriptive of what-is.

But then suppose we say, "'The earth is flat' does not describe what-is." But do we know this? Usual empirical accounts claim that ultimately we must compare what we say with sense data. But sense data have to be interpreted, and this can occur only if there is some original harmony between our sensations and the speech we use to report them. In this respect Heidegger's position resembles an idealist position, minus the metaphysical interpretation of "mind." We do not compare our sentences with things in themselves, but rather with the original disclosures, which are also permeated by language. There is, on this view, no way to get outside language. What we can and must do is compare one level of language with another. Language is the ultimate field of human existence and all that comes into this field—plant, animal, stone, idea. Language is the *Haus des Seins*.

Man, moreover, as we have seen, has a part to play in these original disclosures. This is displayed in a wide manner of ways—for example, scientific discovery, mathematics, poetry, even commerce and technology. The process of letting-come-to light (*Zeigen, Erscheinen lassen, Er-eignen*) uses (*braucht*) man to gain its intelligibility and structure (*ID* 13-34; *USp* 252, 256-67). Even Nature (*physis*) uses man in this regard. What would a tree *be* if there were no one to apprehend it? I suppose Nature "in itself" would be a blind, chaotic striving unless there were some kind of intelligence to give it meaning. But a determinate object has to be *disclosed as* this or that in order to even be what it is. It is a mistake to think of Nature as being in itself composed of trees, animals, stones. Only a naive realist would accept this. We have no evidence that an earthworm knows it's an earthworm and not a grubworm. It probably doesn't make much difference to him. Yet for an object to truly be what it is, it must have an identity *as* this or that. It must appear *as* what it is and must at least in this respect be self-determining. Material nature in itself can at best only comprehend the crudest forms of identity— repetitive patterns. But Heidegger points out that Nature does include within itself its own self-determining power, for it includes man. Man is *physis* par excellence. And mortal speech itself belongs to Nature insofar as it permeates the field of human existence. The unfolding power of *physis* and of language are one and the same power. Heidegger calls our attention to a line of Hölderlin's poetry where he speaks of words arising as flowers (*USp* 206-8).

What kind of an unfolding power does Heidegger have in mind? How do we experience it? Again Heidegger turns to the poets, this time George's poem entitled *Das Wort*. The poem ends with the following couplet:

> "Thus I learned, sadly, to renounce
> No thing may be, where the word breaks off."[21]

To learn renouncement (*Verzicht*) is to acknowledge this harmony spoken of. Negatively, this means that the poet must give up his humanistic orientation and acknowledge that he does not use language but rather language uses him. Positively, it means that he must acknowledge the (usually hidden) power of words over things—that

Kein ding sei wo das Wort gebricht. But this too has a double implication. On the one hand, it appears to point out that words give us a mastery over things. Through words the poet can bring remote things close or close things remote. To do this, he uses names. Names enhance what-is: "These are words [*Worte*], through which what-already-is is made to glisten and blossom and so prevail throughout the entire land as that-which-is-beautiful" (*USp* 225).[22] By using names the poet calls things; he goes beyond the realm of what is "already there," that is, given actuality. The things he calls come to be present, but of course not in the way things are present which lie around us such as a pet, a chair, a sofa, and so on. Yet to Heidegger the presence of these things called by the poet is no less "real." In fact, as was seen in the preceding chapter, the poet's things are more truly *things* than those material objects around us whose presence we scarcely attend to. He asks: "Which presence is higher, that of things-lying-around us or of things-summoned?" (*USp* 21).[23]

In his essay entitled "Das Ding" Heidegger made the point that he does not believe that things are just "there." Things must be given their identity in order to be things in any true sense. Things can only be encountered when they come into our proximity and we attend to them; there are no such things in themselves. Things only come into their own as foci of a complex interplay between the most important cosmic realities: gods, heaven, earth, men. Those things which are *called* to presence have a higher quality of presence, because through being called they are given an identity. Yet we must remember here that the poet is only the instrument of language; thus, it is not the poet who calls but rather language itself calling, and the poet is only successful as long as he hearkens to the call himself. This is where the negative aspect of the renouncement (*Verzicht*) comes in.

The position Heidegger takes here with respect to the language and the presence of things is not as odd as it first seems. For example many philosophers would maintain that objects of (rational) thought are in principle more accessible, more transparent, than material objects. Plato and Hume would not be surprised by this claim, although for very different reasons. Scientists would agree that there is some, hopefully small, discrepancy between a theory and what the theory depicts. Nature, they would say, is to some extent opaque to man. Yet Nature is in another sense the realm of the already-given, that which is not called forth but rather there of its own accord in a chaotic way. Objects of (rational) thought are more transparent to thought, but they either require a deliberate and difficult focus of attention (Plato) or they are somehow themselves human products (Hume). In either case, the object could be said to be called forth whenever it is overtly present. Otherwise it could at best be present in only a hidden manner, perhaps as part of our "unconscious."

Yet, language involves an ensconcement, a withdrawal.[24] This is characterized by the fact that the poet cannot find a name for everything, and this means he cannot bring everything forth in simple clarity and closeness. For example, the essence of language cannot be called forth to be present as a thing (USp 236). The poet can only call forth those things for which names are granted him. Words are granted him but on occasion also withheld. He must hearken to the call of language.

What about words? How are they present? Are they things-which-are? Heidegger claims that "Word and thing are different, if not separate" (USp 192).[25] How are they different? According to Heidegger a word cannot have the exact, same status as a thing and still remain a word. Hence a linguist does not talk about words in any genuine sense but only about the sensible, material vestiges of words —what Heidegger calls word-things (Wörterdinge). The metalinguist also attempts to treat words as objects or things. Neither the linguist nor the metalinguist approach language from the original scene of disclosure. To treat words as things or objects of a metalanguage is to fail to take full account of their dynamics, to impose a static character on them which conceals their dynamic nature. Words must be listened to, and any attempt to talk about them is dependent upon this listening. Words have a special function which sets them apart from things. "The word: the Givent. What then? According to poetic experience and according to the oldest tradition of thought, the word gives: Being" (USp 193).[26] I have coined the word Givent to designate an on-going where the agent and the act are not ontologically separate from one another. Both are defined out of a very complex ongoing and are thus only modally distinguishable within this more complex ongoing. We have seen that Heidegger uses such participial forms as das Gebende to designate such ongoings, which are prior to and more complex and concrete than the simple picture of "somewhats" acting or the "actions" of somewhats. By the word Heidegger does not mean any actual, particular word but rather that which ensconces itself in every word. The kind of abstraction Heidegger is making is somewhat indefinite in character; he has in mind some word but not this or that spoken word. He is talking about a general somewhat which cannot be rigidly specified but is in some way participant in any actual speaking, although itself never spoken. Yet without this unspoken word, all our vocalizations would be sheer noise.

The word is not to be construed as a word-thing which then somehow acts or "signifies" but rather as a kind of on-going in which giving occurs but without any antecedent given. In general Heidegger does not think that there are any simple situations of ready-made agents that can perform actions. This is one of the reasons for his objection to the theory of language as an "activity" of man or to the view of man as simply an ontic being.

The word is not a thing, but neither is it a nothing. One of the cardinal points in Heidegger's thinking is that philosophy cannot hold itself rigidly to the stark alternatives of "it-is" and "it-is-not." Heidegger himself prefers to think in terms of a very special kind of "becoming" or "on-going" which I have described earlier as "ecstatic." By this I mean to indicate that these occurrences (*Entbergen, Verbergen, Anwesen, Abwesen, In-die-Nähe-kommen, Dingen*) relate back to the Region (*Gegend*) within which particular events or things are allowed to become present and take on significance but also allowed to depart and fade away into oblivion. Here Heidegger accepts a dynamic view of Reality but rejects any nominalistic interpretation of it; ecstatic occurrences are generic, enclosing and including all particular, ontic items that come to be present within them. Things which come-to-be have a certain directionality about them and are a function of a point of reference; they must be *here*, in the world, the realm designated not by abstract spatial-temporal factors but by the boundaries of the human, the (self-ensconcing) divine, the earth and the bounded boundlessness of the sky. That is the true realm of human existence for Heidegger—the Di-mension—and it must be the source of measure for what is here (VA 187–204). This cannot be determined abstractly in terms of "being" or "nothing," for these terms lack any kind of a phenomenological indicator whereby we would establish one or the other. To put it another way, for Heidegger, if anything *is*, it must be present. To be is to belong to the open realm. To be meaningful is to belong to the domain of possible human encounter; this is Heidegger's version of the "meaning criterion." Yet the domain of human encounter is concerned with a neutral realm or state in which things neither simply are nor are not but are permeated with a vagueness and openness where the stark alternatives of "it-is" or "it-is-not" cannot apply except on a very limited scale.

We have seen that, for Heidegger, arrival or presence also involves a withdrawal, a reference to a neutral realm or state in which things neither are nor are not; "they" are in this sense ensconced (*ge-borgen*) within the simple foldedness of the twofold, and any presence or arrival involves a dis-ensconcement (*Ent-bergung*) or release from this neutral state. The significance of this was seen to be that all things-present are infected with a radical finitude which foreshadows their possible departure; for their very presence is itself the mark of an ensconcement of the simple fold, the openness of the Region as such. To think correctly about such things we must think in terms of the dynamics of arrival and departure, release/ensconcement, revealing/concealing. There are no things in themselves, but only arrivals and departures as foci and shifts of foci in the quiet interplay of the world-foursome. For example the (occurrence of) blooming is the arrival of the flower. Blooming is a complicated ongoing involving the co-presence of innumerable factors; it is not a simple action done by a thing. The occurrence of blooming is generic because it includes and conditions but is not included or conditioned by all particular flower-bloomings. It is the "form," the "to-bloom" of all particular flower-bloomings. The generic occurrence of "form" is "ecstatic," that is, outside the order of directly encounterable, ordinary phenomena; it never becomes directly present but instead allows flowers to become present and supervenes through these flower-bloomings (*WhD* 133). But the flower bears the mark of an ensconcement, the withholding of the pure ecstatic "to-bloom."

Words too must be viewed in terms of these dynamics of arrival/departure, release/ensconcement, revealing/concealing, rather than as static signs. We can distinguish what arrives (a particular spoken word) from arrival itself, even though the arrival spoken of here is perhaps peculiar to language. Heidegger refers to this primordial arrival as call, oracle, exhortation, Saga (*Ruf*, *Spruch*, *Zuspruch*, *Sage*), which supervenes indirectly through the ontic, spoken word, but itself can never be spoken nor become directly present. It is "ecstatic," giving/withholding, revealing/concealing, unspoken/speaking. The primordial word can never be spoken by man; withdrawing, it speaks to him (*Der Zuspruch*) and through him so that his mortal speech is founded on a listening response. The quality of man's speech is thus dependent upon his attunement with the primordial word.

The word gives, Heidegger tells us, Being (*Sein*). He attempts to clarify this by saying: "A condition is an extant ground of something-which-is. . . . But the word does not ground the thing. The word lets the thing be present as thing [i.e., rather than merely something lying around without an identity]" (*USp* 232).[27] The term "lets" emphasizes here the fact that words do not have causal efficacy. They do not affect the ontic dimension governing the way things produce other things but rather the horizon of presence in which things take on significance as arrivals, momentary endurings, departures. Without words and the ontological dimension, what-is would simply take place blindly without significance. We would be uncognizant of anything; everything would be dissolved into the neutral state of ensconcement, where things have no identity and can neither be said to be nor not to be. The dimension of Being (*Sein*) is the horizon which provides the field of meanings and relationships under which significant encounter becomes possible. Thus the word gives Being (*Sein*) but not things themselves as somewhat existents (*das Seiende*), because by structuring the horizon of presence the word lets a thing be present as what it is.

We can now characterize the relation between words and language (*Sprache*). The ontological word belongs to the realm of the arrival or ecstatic occurrence or unfolding power of language, to which man responds through his mortal speech in the form of words of a particular language. The fact that there are many mortal languages reflects the manifoldness of this unfolding power and the possible ways to respond to it. It also reflects the elusiveness or withholding character of the call, so that the question "What is the right response?" has no simple answer.

Because of the withdrawal inherent in the essence of language, Heidegger calls the ecstatic occurrence of the primordial word *das Geläut der Stille* (the resound of silence) (*USp* 216). Yet silence does not mean the absence of sound but rather refers to the stilling of world and thing. The interplay of the world-foursome is an interplay of silence; it is the ensconced source of the call, where all meanings and distinctions are born apart and merged back together. Because the interplay is one of silence, there can be no rules formulated for it; man is simply drawn into the mysterious, hazardous Region where the interplay goes on. To follow the play of language, which in the everyday context seems so regular and rational, is ultimately, to follow the deadly play of silence itself. For if we ask what grounds our ontic language we are confronted by this silence, the withdrawal of language, and finally the withdrawal of our own question which comes back to us as a fruitless, perhaps even "pseudoquestion," so that many turn away, embarrassed.

Language is, as Parmenides had characterized it, the moving power which unfolds the twofold; language moves the world-poles vis-à-vis one another in their quiet interplay. Language is in a sense an unmoved-mover; it does not itself move but it incites and assembles all motions, all occurrences. Only within the "house" of language can there be an appearance or arrival and a heeding-apprehension thereof; only through language can things be called forth and thereby released from their ensconcement: disensconced. The occurrence of language is self-spectacle par excellence, the occurrence of all possible occurrences: das Er-eignis.

We have seen however, that this interplay necessarily implies the di-stilling of world and thing into residual moments. Thus Heidegger tells us: "Language unfolds its essence as the on-going, self-spectacular inter-cision of world and thing(s)" (USp 30).[28] The out-bearance of world and thing is the mark of an ensconcement. This ensconcement is also reflected in language, namely in the fact that when mortal words are spoken the primordial silence must be breached. The spoken word always involves an ensconcement of the primordial Sage because the former is a reply or response to the latter. A response is always the ensconcement of an original saying, for the response always pushes the latter into the background. Because this is the case, we can become taken up with our own words and forget that they are at best an imperfect, finite response to the primordial summons. When this happens our words become mere names for already-present objects lying indifferently around us. We do not recognize that the objects only appear because they have been so designated by the primordial on-going self-spectacle, that they have been released through a primordial saying which first makes possible our own meaningful speech (USp 262).

Where are words before they are "spoken" or after they have died from our lips? This would ordinarily sound like a category mistake, but in light of the above analysis, Heidegger considers it a meaningful question. It is, oddly enough, precisely because words can cease to be spoken that they can be understood. To take them at face value would be to perceive only noise. We must listen not only to the superficial texture of sounds that come from a person's lips, but to the meaningful, "silent" underlay on which they are predicated. We must listen through the audible sounds to the arrival of the words ensconced in these sounds, as Heidegger would put it. As words die away, they do not fall into nothingness and oblivion; rather, "The fading sound of the word turns back into the soundless, into that out of which it is nourished: The resound of silence . . ." (USp 216).[29]

Using and misusing language

It would seem that there are almost as many different responses to the primordial word as there are mortal speakers. The responses are often contradictory. Suppose someone responded by saying, "There is no language," or, "I am always right and anybody who thinks otherwise should be put to death." Language, left to itself, can be the source of treachery and confusion. The Sophists saw this, as did Socrates. Language would appear to be a blind, pliable medium which can be put to good or bad use. With language we can make the worse argument seem the better, or we can use language to clear up the difficulties of the worse argument.

Heidegger makes very strong claims for the status and power of language. But it is important to note that he does not make these claims except at the proper Region of language, where language—rather than man—speaks, for example in the case of the relinquishment (*Verzicht*) of George. Man has a tendency to respond thoughtlessly to the primordial word—to become so taken up with his own interests that he does not pay careful attention to the silent call and takes the power of language to be his own. It is only when we restrain ourselves and let language speak that language becomes a source of disclosure.

This distinction between "authentic" and "inauthentic" language can be brought into sharper focus by comparing it to Kant's distinction between the heteronomous and autonomous will. Kant's distinction centers around the use of maxims. The vast majority of our actions spring from the heteronomous will and are founded on our desires, inclinations, personal opinions. Conformity to moral law must transcend personal idiosyncracies; we must listen to the voice of rational duty rather than the voices of our own passionate natures. Yet we do not thereby depersonalize ourselves; rather, it is through hearkening to the voice of duty that we first become individuals in any positive sense.

There are problems, however, that leave Heidegger open for severe criticism from other philosophers. How do we distinguish between authentic and inauthentic uses of language? How do we know when we are *listening through* something rather than merely coming up with our own pipe-dreams? Suppose someone says, "Listening not to me but to the *logos* it is well to agree that all things are one." How do we distinguish Heraclitus' saying as "authentic" and worth listening through whereas, "Listening not to me but to the *logos* it is wise to agree that all things are thirteen," is "inauthentic" and not worth the scrutiny of a Heraclitian fragment or a poem by Hölderlin? What could this possibly mean: "Listening not to me but to the *logos* . . . ?" Most philosophers would claim that we need a criterion to distinguish the *logos* from *Heraclitus'* own *logos*. Heraclitus invites us to find *the logos* by relying upon something other than his word, for example, our own careful observations about the way things and their opposites interplay. But Heidegger can offer no such independent route in the case of language since there is no way to circumvent language.

Kant in his moral philosophy was worried by these problems. He tried to set up objective criteria to distinguish authentic maxims from inauthentic ones (categorical and hypothetical imperatives). He saw that this is no intuitive matter and required some sort of objective guarantee. Imperfect as his solution may be, he was sensitive to the problem. But Heidegger gives us no such objective marks. He seems to dogmatically declare that this or that poem, this or that pre-Socratic fragment will serve as an adequate point of departure if we listen to how language speaks through it. Furthermore, there are evidently no formulatable rules to follow in his "art" of interpreting the message of language. How do we know we are really hearing language rather than our own opinions?

Why does Heidegger neglect to provide us with objective criteria? Surely a man of his stature cannot have overlooked such an important problem. What good would it be to search the essence of language if we have no way of locating true instances where this essence arrives? Doesn't this make all points of departure spurious and arbitrary? On the other hand, in order to formulate such criteria as are here being called for, we would have to have insight into the nature of language. There is indeed a circle, and unless we are already inside it—unless language has already spoken-to us—there would be no way to begin its traversal. In *Sein und Zeit*, Heidegger discusses this "hermeneutic" circle and claims that the circle is not vicious, because the only way the human understanding can really work is by moving within this circle and tracing out the paths it embraces. Only by making the attempt to complete the circle can we find out anything about the circle itself. As Mr. Voelkel points out, we are always in the circle because we are part of *Ereignis*, the self-spectacle which "never sinks away." Thus it is futile to ask the question, how do we know we are really in the circle. To suppose we are not in the circle is to reject the claim that *Ereignis* has on us. This would mean turning away from our own destiny (*Geschick*).[30] Here Heidegger shows a profound distrust of formal criteria in themselves. Searches for formal criteria are often predicated on a panicked desire for control and certainty, as though one could learn about the circle without having to take the pains of traversing it. This would mean backing off from the leap.

"Authentic" sayings can appear contradictory, as do those of Heracleitus and Parmenides; but instead of trying to reconcile contradictions and opposites and insisting that they aren't really there, Heidegger attempts to show a kind of harmony underlying these various sayings, for example, those of Heracleitus and Parmenides. Authentic sayings do not need to be rigidly consistent with one another; but they must in some way harmonize with one another, while at the same time reflecting the diversity and manifoldness that Heidegger takes to be characteristic of the supervention of the open realm and the drift within it. Some of these sayings may indeed be "false"; error is after all a large part of man's existence. The word is ambiguous and elusive; it discloses but it also conceals. The necessary unity of revealing/concealing shows Heidegger's view of human existence to be a tragic one in the classic sense. There is no foolproof method of intelligence, since intelligence itself oscillates between, bears-apart, and gathers back up the poles of revealing/concealing.

If we attempt to traverse the circle, we can often see that certain sayings reflect more transparently back to their own primordial context or historical epoch (*Seinsgeschick*). It is virtually impossible to write Heracleitian fragments today and pass them off as philosophy. This is again due to the drift of the open realm. Since this drift is again part of *Ereignis*, our saying today must appropriate itself to this drift. On the other hand, when confronted with a saying, we as listeners must listen through what is said in order to attempt to gather in what is on both sides of the drift. Thus the authenticity of a saying is as much a function of the listeners as is its source. There is no such thing as an authentic saying in itself, and it is this fact that makes it impossible to formulate formal criteria for authentic and inauthentic language. In other words it is impossible for us to describe the structure of the circle except by the very act of going around it. Thus the distinction between authentic and inauthentic language appears to be a relative one. All sayings are authentic to the degree that they permit the arrival of the primordial word in the mode of ensconcement. All sayings fall short of bringing the word to words because of the ensconcement; and to the degree that the ensconcement becomes complete, the saying becomes a mere matter of words, inauthentic.

Then what about ordinary language? Surely Heidegger cannot say that a housewife is responding to the silent call of the word when she asks her grocer for a pound of lentils. Has he overlooked the obvious? By no means. He tells us: "Authentic poetry is never merely a higher plane of ordinary language. Rather, on the contrary, everyday speech is a forgotten and worn-out, over-worked poem, out of which the call is scarcely audible any more" (USp 31).[31] In everyday language the call is scarcely audible, but it is there. Perhaps the involved housewife does not attend to it, but an artist might. Some of the best art has as its subject matter very ordinary things and scenes which then are suddenly transfigured in poetic splendor. The everyday is not a world unto itself, it is a system of lax habits of inattentiveness into which we fall when we become seduced by the glittering things around us and fail to remain open to the silent call which allows them to be present. The result is a depreciation of the things themselves, since they are no longer looked upon as having any sort of cosmic significance. But then on the other hand it is of the very nature of the call to elude men, so that they become all the more easily inattentive to it.

Heidegger is not casting aspersions on those who are firmly entrenched in the everyday; for being "authentic" or "inauthentic" is not a simple act of the will. It is rather like listening for a certain theme in a quartet; listening involves waiting. We cannot listen to a theme in a quartet if that quartet is not playing. Similarly, there are factors—mysterious, hidden ones—upon which our response to the silent call depends. We are in a way at the mercy of language; as Heidegger puts it, even though we fancy ourselves as masters of language, actually language is a master of man (VA 146). The example of this which Heidegger treats in greatest detail is the poet George's renouncement (Verzicht).

Thus language itself is of cosmic significance for Heidegger. It is not merely a human phenomenon. In this sense his objection to a utilitarian instrumental outlook on language amounts to saying that this outlook is short-sighted. It involves a failure to look deeply enough into the dimension in which we most truly dwell. This dimension is much larger and deeper than "actuality," for it must include an interplay of all modes of presences and absences. It is the polarized Region in which the four most important cosmic realities—earth, gods, sky, and men—interplay and reflect themselves into one another (VA 187-204). This dimension is the dimension of language itself.

Whereas most philosophers would be primarily concerned with distinguishing the things within the realm itself and thus attempting to gain some mastery over them, Heidegger attempts to understand the realm. There is a certain piety to his thought. We should attempt to master certain things in order to make worthwhile things; but technology and production are blind without a searching inquiry into the essence of man. Similarly, we should use language to help us think about, disclose and meditate worthwhile things. Heidegger is not arguing categorically against the appropriation of language *per se* but rather against thoughtless, unappreciative appropriations of it. The latter tend to be insensitive to the source of language, but the former need not be. This ability to use, appreciate the cosmic source of, and not fall slave to our own appropriations is an example of Heidegger's *Gelassenheit* (detachment, acquiescence).

What follows from all this about Heidegger's own use of language? If words are translucent hints rather than transparent designators of invariants, language will always be a major problem for philosophy. It will be a mistake to hold ourselves to the letter or "content" of any text, as though the text were a cup holding water, so that the one could be ignored in dealing with the other. Rather, we must pay attention to the language of the text as dynamically unfolding a path on which to travel. Language and content not only cannot be separated, but they cannot be properly distinguished. Content is itself linguistic in nature. Thus we must pay attention to the language underlying the text; we must try afresh each time to listen through the text to the still underlay on which it is founded. This may sometimes require finding new ways to say things. The philosopher must, in this respect, imitate the example of the poet. He must develop an appreciation for his medium—not merely an aesthetic appreciation, but a pious one which reflects the cosmic significance of the medium. Rather than forcing something into a standard style of expression—

rather than listening uncritically to convention and tradition, the philosopher must learn to listen through these and to follow the play of language itself.

In discussing the criticisms of Heidegger, I do not mean to indicate that Heidegger has a completely adequate reply in store for them. He has a reply, and he is evidently prepared to trace out its tragic consequences. But in doing so he leaves himself open to the undesirable features of any determinist position; by preoccupying ourselves with the cosmic significance of something we tend to lose our feel for open perspectives on the things around us. Versenyi[32] has commented that in Heidegger's own writings there is a growing insensitivity to the open receptiveness which Heidegger himself thinks is necessary to philosophic thought. He tends to adhere to his own expressions, to use them over and over without attempting to clarify them. Examples are his use of *Ge-Stell, Ereignis, Gegnet, Geschick.* Has Heidegger failed to listen through his own words? Can he make the distinction, "Listening not to me but to the *logos* . . ."? In refusing to lay down formal criteria for such a distinction, Heidegger gives us no recourse but to follow his "art" and observe his example. But this lays upon him the responsibility to provide examples beyond reproach. If he does not, he undercuts his own position, as though a piano teacher said, "Play it like this," and then fumbled.

But suppose Heidegger were successful in making the distinction between listening *to* and listening *through?* Would this not tend to eliminate the content of our sayings altogether? If we find hidden harmonies among even the most violent contrasts of ideas, it would seem to mean that it really doesn't matter what we say and it doesn't matter what we hear as long as we are able to listen through it to the silent call on which it is founded. But such a view has the effect of completely obscuring the nature of language or perhaps eliminating it altogether as a mode of any kind of selective disclosure. We would be back to a variation on Parmenides which would run like this: No matter what we say and how much we talk, only one thing really gets *said,* and it says itself.

To put the problem another way, if Heidegger is going to distinguish between an ontic, spoken level of language and a primordial, silent call, should he not somehow positively clarify this distinction, either by exhibiting a mechanism whereby one comes over into the other, or by a formal criterion whereby one is exemplified by the other? A man cannot call his spoken words a response to a call if he is completely left in the dark as to the nature of what he is responding

Language and Ontological Difference   187

to. In what sense is this really a response rather than a lonely cry from out of the darkness? If this cannot be explained, doesn't Heidegger's twofold portrayal of language also fail in the end to illuminate the nature of language?

But here again we betray ourselves. We want certainty and clarification. Perhaps the hardest thing to face in Heidegger's thought is its sober willingness to entertain the possibility that these may be only one side of a tragic revealing/concealing. Perhaps we lack Heidegger's *Gelassenheit*, his detached but compelling openness for visions beyond the thing-, people-, and place-world and all its grounds and reasons. Perhaps we lack his willingness to face up to a philosophical impasse without fear of embarrassment.[33] Could it be that the very nature of man's response is that it is all too often doomed to be a cry from out of the darkness, the black night into which shadows retreat when the world becomes "clarified" with the glittering brilliance of scattered particulars and all their grounds and reasons; when everything seems evident and obvious and in no need of searching philosophical questioning?

VII

The Continuing Mystery of Ontological Difference

Unwirtbar wär es, ohne Weile;
Was aber jener tuet, der Strom,
Weiss niemand.
 (*Hölderlin*)

The ontological difference remains elusive regardless of philosophic attempts to bring it to expression. This is necessarily so, since the ontological difference is an unfolding/folding whose simple foldedness withdraws as the things-which-are come to be present. Elusiveness is often associated with the abstract and remote, the theoretical, that which is speculated upon with a kind of idle curiosity. I have attempted to show, however, that Heidegger's interest in the ontological difference is more than an abstract, theoretical one. The implications of the ontological difference for his thinking are so deep and far-reaching that they penetrate into an inner unity that is prior to any distinction between "theoretical" and "practical." This inner unity can be glimpsed in the poetic outlook, where concern with things is not based on their crude ability-to-do-work but on a more comprehensive context which incorporates this functionality of things into a living presence (*An-wesen*) where both world and thing are vis-à-vis one another. Here, in the Region marked out by the polarity of world and thing, is where authentic man most truly dwells. Such dwelling is at once more "theoretical" than ordinary theory and more "practical" than ordinary practice. Dwelling is poetic concern, which directs itself not merely to things lying-around (*das blosse Seiende*) but to things-in-their-presence, the twofold what-is/be-ing (*Anwesendes/anwesend, Seiendes/seiend*).

Thus the ontological difference is not an object of thought that can be conceptually grasped and categorized in any adequate manner.

Mystery of Ontological Difference 189

It is not a problem for ontology to solve. Rather, the twofold, the di-ference, first calls forth thinking as an authentic response to the primordial injunction (*Geheiss*). It is to-be-thought (*das Zudenkende*), that which always lies out ahead of thought and draws us out and onward. The ontological difference *underlies* and thus remains ensconced within our searches for grounds and justifications; the Difference itself cannot be grounded or justified. Since we are drawn out into the open realm and set adrift by its drift, we do experience the Difference without needing to schematically represent it as an object of thought and attempting to ground or justify it. The thinking that would *openly* follow the primordial injunction is in fact forced to leap away from conventional certitude and groundings and to face the hazardous openness of the Region into which we are drawn. That we are always subject to this hazardous openness is for Heidegger shown by the way in which dread, guilt, death may steal over us unexpectedly, even in the middle of our most brilliant groundings and certifyings.

The most important implications involved in a recognition or paying-heed to the Difference are, I think, these: (1) Man is not just another thing lying-around. Man is in essence twofold, for he is openly related both to the things-which-are and also to the open Region of self-ensconcing presence within which they reside and into which they can at any moment disappear. Man is essentially a poetic being, who is concerned with things and also with the Region which grants them their presence. Although man can fail to openly acknowledge his twofold status, he cannot escape from this designation. Man's essence is called forth to this threshold where things come to be present and where the Region which grants them their presence ensconces itself in uncanny, empty openness. (2) Thus the self-ensconcing disensconcement creates a gap between the openness of the open realm and the things-that-are-in-the-open. But because the openness itself is self-ensconcing, or self-withdrawing, there is a drift to the open realm itself. Called forth into this openness, the essence of man is at the mercy of this drift and is set on an errant, wandering course among the things-that-are. This drift is the drift of history itself. (3) The drift has the dual effect of estranging man's essence from the original Region and also uprooting the things-that-are from their originative source so that they appear as commonplace, mere things-lying-around. There is thus an inner harmony (*Ereignis*) between the uprooted mode of everydayness and the uprooted mode of historical estrangement. (4) Because truth is disensconcing/ensconcing, authentic thinking cannot proceed one-sidedly towards the goal of clarifying all things. Rather, authentic thinking must assume the character of letting-be, which respects and acknowledges ensconcings and withdrawals as well as presentations and graspings of things-that-are. Mystery is real and thinking has a kind of piety (*Frömmigkeit, An-dacht*) to it but is not thereby made idly superstitious. (5) The narrower notion of object gives way to a "higher presence" of things which are *called forth and let be* (*hermeneuein*) rather than stumbled upon passively as merely there, already lying around us or blindly appropriated to human contrivance and convenience (*apophansis*). This has the result of leveling out or at least relativizing the importance of some traditional ontic distinctions, such as material-formal, actual-possible, physical-mental, right-wrong. Yet this does not mean that all things, as seen from the standpoint of the Region, are on the same gray level. These distinctions are replaced with more dynamic ones which center around the locus (*Ort*) of human existence, the threshold of the Difference as unfolding/folding: present-absent, proximate-remote, arrival-withdrawal. (6) Corresponding to the "higher presence" of things is a "higher unity" of experience. The inner attunement of *phainesthai* and *logos* involves a broadening of the domain of possible experience to include a kind of "seeing" (*Er-blicken*) and hearing (*Er-hören*) which are attuned to the primordial word or injunction itself (*Geheiss, Zuspruch, Sage*) rather than merely to the things disensconced within that Region. This attunement lets the things come forth, calls them, and is thus *hermeneuein*: announcing and letting-come-forth-to-be-seen. But man can only partake of such experience when he brings himself to dwell properly within the Region itself. This is accomplished by the leap, the step backward, the "Reduction" (*Sprung, Schritt zurück*), which overcomes the tyranny of the "it-is." (7) The Difference permeates the nature of language. Hence language cannot be adequately treated as a system of conventional symbols. Language is a living presence which arrives and withdraws and is thus also twofold. Language is *hermeneuein*: announcing and letting-come-forth. But *hermeneuein* itself occurs, indeed in a much more essential way than any other event. *Hermeneuein* is the occurrence of all occurrences, which Heidegger calls *Er-eignis*. The occurrence and on-going self-spectacle (*Er-eignis*) of this announcing itself is what must now be acknowledged by the philosophic thinker. This is the occurrence of language itself: the silent interplay involving the world-foursome and a breach of this silence as mortals utter words in their response to the primordial call ensconced within this interplay. Yet rarely do men listen through their spoken words to the silent call which evokes them as speech rather than merely vocal noise.

Some critics of Heidegger have claimed that his ontological difference is based on a "linguistic confusion" or bad etymology. The implication is that his thought would not have taken the course it has, had Heidegger not fallen victim to this "confusion." Throughout his work Heidegger has emphasized the need to follow the clues tucked within language itself. In the last analysis, he argues, all phenomenology is hermeneutic and must therefore listen to the message of language itself. To talk of linguistic confusions is either to presuppose some nondiscursive "fact in itself" that lies beyond or behind our discursive awareness (an unprovable assumption) or to draw arbitrary lines as regards the meaningfulness of language. The test of a meaning criterion must surely be based on the degree to which its application proves fruitful in illuminating the nature of human existence and the kind of world therein implied. It cannot be denied that Heidegger's interpretation of man, world and thing is penetrating and profound.

The Difference is not based a priori on a mere "linguistic possibility." It is not, strictly speaking, "based on" anything. Rather, the Difference is named as a result of an encounter with and through language—an experience of the disensconcing withdrawal, silently spoken-to (*zugesprochen*) us with and through each word, provided we are attentive in the right way as were, for example, Heraclitus, Hölderlin and George. Above all, the Difference is not something arrived at by a generalization of particular withdrawals. Heidegger agrees with Kant that an inductive procedure will never lead us to the kind of transcendental principles supervening on the ontological level, which grounds and makes possible the various ontic particulars. If we are not already in some sense "in" and at the scene of the disensconcing withdrawal, no procedure which starts with the pure ontic and builds on it will ever lead us to such an encounter. As Heidegger puts it: "No path of thought, not even that of metaphysics, goes out from man and from there over into the realm of Being or conversely from Being out and then back to man. Rather, every path of thought always *moves within* the entire affinitive relationship between Being and man; otherwise there is no thinking" (WhD 74).[1] This to Heidegger is the original meaning of the *a priori*: that to which our essence is already related in an original and fundamental way. The problem for philosophy is not to discover something new, but rather to reorient man's attention to that which always from the very beginning surrounds him and escapes his notice (SvGr 154). The problem for thinking today is to come to terms with the estrangement between man, world and thing. By realizing that Being cannot be dealt with as something-which-is, man can hold himself open to a deeper and farther-reaching articulation of his existence.

As for Heidegger's etymologies, most critics overlook the fact that Heidegger is interested in reaching into the *implicit* meanings, the hidden un-thought (*Un-gedachtes*), the un-spoken (*Un-gesprochenes*), which remains ensconced within the words in question and yet in the course of time can be discerned by those who take the step backward and attend to the fated message of the unfolding/folding (*das Seinsgeschick*). Thus, any criticism of Heidegger based on "what the Greeks really meant" remains only with the explicit, revealed aspects of language; such analysis operates on a superficial level as far as Heidegger is concerned. Because the "descriptive" approaches remain only with the revealed aspects of language, there can be no evaluation of the degree to which a culture thinks out the implications of its relationship with language—myth, poetry, philosophy, information. Descriptive approaches that shy away from questions of evaluation would investigate every culture alike. Heidegger believes that there are deep-rooted cultural differences, and that these are reflected in the way a culture is related to its language and in its interpretation of man, world, thing. Some traditions are full, rich, poetic; others are empty, barren, superficially functional.

The Difference is not meant to "explain" anything in the sense of grounding particular ontic facts. Rather, as the primordial injunction, it abides as a leading principle which draws us along a path, pointing into the boundlessness of the open Region. The ontological difference underlies but also unmasks the search for grounds and certainty. The success of such attempts depends upon the open Region shining forth as absolute Ground. What happens if this light is discovered not to shine? The twentieth century is faced with this problem.

It is widely accepted today that the light which apparently illuminated the metaphysical tradition does not shine. Some think that this was the fault of metaphysicians themselves. It is argued that they failed, for example, to pinpoint the domain of meaningful inquiry and were thus guilty of posing pseudoquestions and then giving pseudoanswers to the pseudoquestions. Thus one reaction to the decline of metaphysics has been to narrow the domain of legitimate

Mystery of Ontological Difference 193

philosophical inquiry, for example, by drawing a sharp distinction between a *priori* (but factually empty) truths and material (and necessarily a *posteriori*) truths. Heidegger, on the other hand, feels that the light and its eclipse both belong to the destiny of the unfolding/folding itself. He seeks to widen the domain of philosophical thinking by reaching back into the soil which nurtured that light, namely back into the Region, the disensconcing withdrawal. To do this, he develops a broader notion of encounter, as has already been noted; by concentrating on the dynamics which underlie and make possible any object presentation, he is able to incorporate absences into experience as well as presences. Moreover, by abandoning the primacy of empirical sight in favor of the primordial unity between *phainesthai* and *logos* we are able to encounter the "higher presence" of which Heidegger speaks—the presence of those things called forth and attended to rather than those lying around us in indifference. The sense of inquiry must be transformed when dealing on this level; for if the things are called forth, then there is no need to "verify" whether they are really there. The problems that arise are of another character, such as that of overcoming the estrangement between man and those things lying-around him, and also the problem of adequately responding to the primordial call which draws us out into the open Region. For, despite their technical precision, the scientific philosophies of the modern and contemporary eras seem to be symptomatic of a deep alienation of spirit; there is an unspeakable chasm between man, world and thing which lurks within the seemingly harmless theoretical problem of the Cartesian dualism. And nowadays, many of our deepest spiritual problems are relegated to the status of the pseudostatement.

Heidegger would agree that the ontological difference cannot be talked *about* in the way questions about ontic facts can. But why should ontic questions be taken as the paradigm of excellence? They give us answers which are often "clear" and "verifiable," but they are also often piecemeal and superficial. They encourage man to "think little" and close off his true Dimension. We have, for example, little difficulty in explaining at the ontic level how a rose blooms; but *that there is a rose*, this ought to occasion wonder. The suggestion that the question "Why is there a rose?" is a pseudoquestion implies that our wonder is irrational, a mere thing of feeling. To live by a philosophy incorporating such suggestions would be to live a life of indifference, where wonder and value would be mere feelings like tastes and smells. Wonder, to Heidegger, is no less rational than a mathematical equation and indeed the former may do much more in helping man achieve a closeness with world and thing. Wonder opens us up, allows us to be responsive to the openness of the open Region. Of course, with wonder comes also curiosity, which can lead us to preoccupation with novelty and fad as we focus on the new things disclosed rather than the Region itself. The inner unity between wonder and curiosity is a reflection of the inner unity which pervades the disensconcing withdrawal and releases forth the colorful multiplicity of things as *ensconced* termini of the Region.

This puts us in a position to confront the criticism sometimes offered that Heidegger loses sight of the ontic, that he loses contact with the things-which-are. Before such a criticism can be meaningful, we must ask, what is it to have closeness with things? Do the Positivists have the requisite closeness? What about the mechanic or the engineer? The aesthete? Heidegger takes the position that spatial and temporal adjacency are not enough to define true closeness; rather a kind of intimacy, a vis-à-vis-ness (*Gegen-einander-über*) is needed in which things are allowed to shine forth and thrive (*USp* 210–12). Do we find this kind of closeness in Positivism or in scientific or technological research? We surely cannot equate precision and control with proper concern and closeness. I do not believe a proper criticism of Heidegger's treatment of the ontic can be formulated until we can answer the question, what would proper concern with the ontic involve? If we cannot answer this question, how can we determine whether Heidegger allows sufficiently for the ontic?

It is true, I think, that Heidegger's concern with the openness of the Region leads to a leveling out of ontic distinctions such as "possible" and "actual," "physical" and "mental," "optimistic" and "pessimistic," and so on. Yet this does not mean that all things lose their significance, as has already been noted. Science is a much more insidious leveler, in that it applies the same basic method indiscriminately to a wide variety of ontic items. A descriptive anthropologist will spend as much time with a barren culture as with one which is rich and profound in its interpretation of the human spirit. A literary scholar today might divide his time up equally among the monumental and the second-rate works; indeed, he may spend more time on the latter because of greater possibilities for publications. It is exactly this kind of indifference to things which Heidegger is trying to overcome by what he calls the true closeness with things—that closeness dependent upon the higher sense of presence, being-called-forth, rather than merely lying around us. The proper element within which to encounter thing as thing was seen to be world: the dynamic, ecstatic interplay of earth and heaven, mortals and gods. Without this context, a thing can only be stumbled upon indifferently as a mere object lying around, a piece of junk.

What evidence is there for Heidegger's belief that man is "estranged" from the things around him and that this is really a problem? Don't most philosophers see the primary problem as one of estrangement between man and man rather than between man, world and thing? Yet what is it that divides men? Isn't it primarily the fact that they can't agree on this or that? This lack of agreement, as Heidegger sees it, is due to a prior estrangement between man and his Region with its poles of world and thing. We are all for the most part uprooted from the Region and its primordial injunction. We are cast out, disensconced, among the things; nevertheless, we cannot take things as they lie around us, for we are also exposed to the open presence within which they arise. Thus man attempts to know the things around him and to ground them; giving ground to things is an attempt to overcome the hiatus between man and thing. Truth has classically been defined as an agreement (harmony) between what man says and what is present. In truth there should be a kind of closeness between man and thing, and therefore also an agreement between man and man. Error and falsehood involve a kind of alienation of man to his Region; he has fallen out of this attunement, he is estranged from the primordial call. Thus science and technology as well as religion and poetry are born out of the attempt to overcome the estrangement between man and thing (*EiM* 90, 105). The paradoxical thing about science and technology is that they seem to estrange man even more, and thus there is a kind of vertigo or frenzy present in the way that research and production are carried on today (*SvGr* 200–3). That Heidegger connects the idea of ground-seeking with that of attaining closeness with world and thing is already seen in the concluding paragraph of his early essay *Vom Wesen des Grundes*: "And so man, as existing transcendence, over-abounding in possibilities, is an essence who dwells in the distance. Only through primordial distant horizons, which he forms for himself in his transcendence over all-that-is, can the true closeness to the things about him arise. And only if he is able to attune his ear to these regions remote, can he bring to fruition for himself the reply of those who are together with him—those in whose presence the I can be suspended, in order that he may be as an authentic Self" (*WG* 54).[2] It is impossible here to do justice to the richness of this passage; nevertheless, I will single out two points. The first is that in this essay Heidegger had sought the meaning of ground in the phenomenon of transcendence and the return to facticity therein implied, similar to the situation described later in *Bauen Wohnen Denken* in terms of man looking up to the heavens and then back to the earth and its everyday familiarity. The implication is that ground-giving and ground-seeking are not autonomous, self-grounded activities. Rather, there is a higher kind of closeness that must be won if these groundings themselves are to prove fruitful and not end in despair and confusion. I do not believe that this view leads Heidegger to an "irresponsible mysticism"; rather, he is pointing up a danger in the now existing un-context of strangeness and blind power which seeks to order and justify everything and is more sensitive to the order of justification and control than to the things present and their presence. The second point worth singling out for our purposes is that meaningful encounter between man and man is only possible within such a context of closeness of world and thing where the ear is attuned to the distant regions. The kind of contact that men have with each other without this attunement is petty and egocentric. The reason there are "ethical problems" is because men become uprooted from their Dimension, wherein a closeness of world and thing would possibly prevail. Thus, any attempt to do ethics without first attacking the prior problem of man's relation to world and thing is doomed to failure and superficiality as Heidegger sees it. This is exemplified in his discussion of Nietzsche's "highest hope" for man, that he be delivered from the spirit of vengeance (*Rache*). Heidegger argues that this is not a "moral problem" but rather involves a disorientation to the essence of time (*WhD* 33–40).

Thus some critics of Heidegger have been led to feel that his philosophy shows little concern for problems of value. However, merely because a philosopher does not write an ethics or an aesthetics or couch problems in terms of traditional value terminology does not mean at all that he is indifferent to the original, prephilosophic situations which generate the formulation of value problems. In fact, many of the so-called value philosophies have lost contact with these original situations, and their discussions seem abstruse and theoretical in the negative sense. Heidegger is unwilling to entrust such problems to disciplines which he considers estranged from the incipient unfolding/folding. The very disciplines of ethics and aesthetics assume usually that one can categorically establish value criteria. Heidegger cannot admit to this assumption, because it is counter to his notion that each truth involves an ensconcement and hence cannot be measured in itself. Furthermore, he questions the very validity of man's measuring himself in categorical terms; he points out the need to appeal to some wider kind of measure in dealing with problems of human existence. He raises the question of the ontological status of human actions and the relation of these actions to the human essence itself. He strikes right at the heart of the problems of measure and value in a way that no traditional ethics or aesthetics has done.

The unfolding/folding, in the guise of a primordial summons, calls forth the essence of man. We respond to this call by thinking, by speaking and paying heed. The primordial summons could be regarded as a kind of "ontological imperative" to which we must respond. If all men could pay explicit heed to this summons and thereby relinquish their claim to order and rule the other beings and entities, all problems of human strife and conflict would probably dissolve into a kind of acquiescent harmony. Each man would regard others with a kind of dignity. However, the primordial summons dissembles itself to man, and he becomes the victim of confusion and opinion. Men are turned away from the primordial summons, and discord, suspicion, hate, discontent are born. The idealistic notion of harmony among men dissolves accordingly on a tragic note; for a harmony which is hidden and dissembled cannot be effective as a principle upon which men can base their encounters with one another. In the end we are thus left with the need for a practicing ethics —a point which Heidegger has not really stressed.

Heidegger is not a "man of action." His concern for men is more of a heeding. He would not lead men to a course of ethical action or self-determination. Yet he is concerned with helping those who dare to venture along the way of the thinker in order to find their own essence. Heidegger does not think of his mission as that of helping all men; for not all are called to go upon the long, lonely way of the thinker. This kind of selective concern is in keeping with Heidegger's dissociation of the principle of plenitude from the unfolding/folding self-designation.

The "ontological imperative" forms the basis upon which any ethics or system of value is possible. Yet the way in which the primordial summons becomes manifest to men is so dissembled and so ambiguous that men are brought into disagreement. Nevertheless, I think Heidegger tries to restore to man a kind of ontological dignity by making him subservient to something more noble and fundamental than the world of categorical entities. Without such a concept of dignity, value dissembles itself into a scattered plurality of provisionary objects of a momentary, blind striving. Values are then left devoid of any final significance and are looked upon more as crutches than as marks of strength.

What about the criticism that Heidegger has left philosophy behind and that he advocates merely waiting? This interpretation arises when one separates Heidegger's thought into "phases" or "aspects" that develop in linear fashion and can be treated independently of one another. I believe this separation can be overdone, especially since Heidegger himself continually connects the idea of an "arrival" with "what has truly been." Heidegger's own work is an example of waiting. Waiting does not leave anything behind; rather it is research and progressive discovery which leave things behind and are always on the lookout for something new and different. Waiting is anything but a passivity or form of dispair; waiting, the supremely careful act of heeding, involves watching, and in order to watch, man must survey his Di-mension. Thinking is a form of waiting or watching, involving the laying down of a path and then dwelling upon that path. Thinking is never a simple exercise that man can perform in a vacuum or mechanically engage in according to some formal rules. Building a path involves gaining a way of access to the proper Region of disclosure. Path-building is necessarily "speculative" in character, for it must make its own way of access as well as stay loyal to the path it lays down. The situation of the thinker is in this respect somewhat like the situation of the artist, who must find his own way of proceeding and then carry it out precisely by remaining loyal to it. In neither case can path-building and path-traversal be severed from one another. To traverse a path we must remain loyal to it and not become sidetracked; this is often tedious, for we must dwell upon the path. The man who is overly anxious to go beyond is all too often impatient and thoughtless, for he overlooks the significance of what is most proximate to him. He is not really *on* a path at all; he is like the mountain hiker who only wants to reach the summit and is oblivious to all the quiet beauties and dangers which lie on the path *underway*. It belongs to man's essence to be underway, *at* the scene of the unfolding/folding but *underway* to the simple fold. As the being who dwells in the Between—the Di-mension under the heavens and upon the earth—man is always underway. This is to some extent reflected in the way in which so much traditional philosophizing makes use of theological principles. Yet Heidegger considers it a mistake to view being underway as a means terminating in some end which we can latch onto and then leave the path behind. The path underway can only be suspended (death), never run through like a series of points. Being underway is meaningful in itself if properly reflected upon; this is the meaning of Gelassenheit: that we acquiesce into the silent interplay and take up dwelling openly at the threshold where world and thing are brought intimately vis-à-vis one another (Er-eignis) and yet also borne apart (Austrag).

The ontological difference demands that we hold ourselves open to the Mystery, the simple fold. One interpreter of Heidegger has criticized him by claiming that Heidegger thus attempts to respond to that which cannot be responded to—what is totally other to man. He claims that the only response appropriate in such a case would be total silence. But for Heidegger it is not Mystery per se that man responds to; rather, he responds to the unfolding/folding as it calls man forth while at the same time withdrawing in its simple foldedness. Man is not juxtaposed to Mystery per se; both are part of the quiet interplay of the world-foursome. However, as Heidegger sees it, we cannot avoid Mystery either; the attempt to turn away from it leads to false certainties which are bound to collapse as dread, guilt, and death steal upon us and we are caught unprepared for their sudden arrival. All men must indirectly face Mystery in some way; the question is not *whether* we respond but how openly we respond to the incipient unfolding/folding and its Mystery. And even this is not solely up to man as some kind of timeless, autonomous agent but also depends on how the thinker is called forth, what is in the open realm at that time, and how that openness disensconces and withdraws. For Aristotle there were dynamic presences which were what they were at any one time but nevertheless came to be and passed away. Yet they had a continuity which was provided for by the fact that the *physis*, whereby they came into being, did not itself come to be but rather supervened as a principle (*arche*). Thus the problem of the ontological difference is ensconced within the relation between *physis* and *ousia*, between letting-come-to-be and what is present. But within *ousia* itself is also a twofoldedness which hearkens back to the Difference. For Kant there were objects whose presence was grounded in something which did not appear as such but made possible the appearance of what does appear. For him the ontological difference is ensconced within the (also hidden) unity of the a priori intuitions space and time, the thing-in-itself, and the noumenon—a unity which Kant did not explicitly see but must have glimpsed in a shadowy manner as witnessed in his discussion of the schematism.

For Hegel there was absolute presence itself, in whose light all appearances could be illuminated, while they in turn precipitated or condensed the otherwise vacuous presence of absolute presence itself.

The rose in the meadow is grounded. Whatever happens ontically to the rose can be explained—too much sun, too much water, and so on. Yet we cannot explain *why* the rose is; its very presence remains a mystery. Heidegger quotes Angelus Silesius, a mystic who lived during the late sixteenth century:

> The rose is without a why; it blooms because (while) it blooms. It does not watch itself, asks not if someone sees it.
> (SvGr 69)³

Why the rose blooms is our problem, not the rose's. The "why" arises because of an estrangement between us and the rose and also between us and the ground upon which the rose grows. We are uprooted from that ground. Because the "why" is predicated on an estrangement, any attempt to explain it in terms of this or that cause or condition will always reflect this estrangement. This may be one reason why the rose gets lost when the microbiologist, the biophysicist, the biochemist or even the florist produce their laws and explanations and criteria for prize-winning specimens. The rose asks no one if he sees it, understands it or finds it beautiful.

Heidegger claims that a hidden implication of this poem is that man must in his own way become like the rose—without a why—if he is to be essential man. He must not attempt to find meaning in his life according to some constructed preordained plan which categorically orders all things; but neither must he turn away from grounds altogether and develop an impious and uprooted empiricism. The rose is without a "why" but it is not without a "while," a region of disclosure. It is within this "while" that man also finds himself; within this "while," the "Region," lies the key to overcoming the estrangement from the rose and the Region. Heidegger is not advocating that we throw reason to the winds and become like the rose in the sense of trying to live like children of nature. Rather, we must let ourselves belong to the same Region which lets the rose flower forth—*physis*, Be-ing, unfolding/folding. Man, open to the openness of his Region, flowers forth in his thinking and his poetry; in responding authentically to the primordial injunction, man lets "words arise like flowers" (USp 206–8).

To achieve a closeness with the rose and to let ourselves openly belong to the while we must heed the silent interplay of the world-foursome which stills the rose and lets it be as it is. The rose is a mark of ensconcement, the simple fold, but also a mark of disensconcement, the unfolding of the silent interplay. The rose carries within it the mystery of being—the simple mystery that the rose *is*. But we can only attain a closeness to it if we let Mystery be—the quiet presence of the rose, the silent withdrawal of the openness of the open realm—to make room for the rose. In the withdrawal of this openness lurks the hidden divinity—close, where the rose is, yet remote, withdrawn, into the clouded boundlessness of the openness itself. This twofoldedness of presence and things-present is for Heidegger that alone which is truly worthy of thought—not to be grasped as an object and then forgotten about but rather continually returned to and dwelt upon, underway, while in the Region, in response to the primordial, silent call which resides within the twofold itself. A simple rose ensconces cosmic secrets. Therein lies its quiet beauty—a beauty which is higher and deeper than what the sensuous aesthete feels, and evident only to those who openly respond to the primordial call, those who are open to the Difference itself.

For Heidegger, then, authentic philosophic thinking must follow the twofold, disensconcing withdrawal, rather than remain with what-is. Thus a philosophy cannot be justified in terms of what-is; the usual questions of verification and validation no longer apply since they are formulated as categorical standards for what-is. Neither is thinking a kind of problem-solving, a means to an end. Rather, it is meditative, dwelling underway, and calling forth the presence of things while underway. In this respect thought is akin to poetry. We cannot read a poem by rushing through to complete it, treating the last line as some kind of goal and the preceding ones as obstacles to be run through in achieving the finish. To properly read good poetry we must dwell and remain with each word, each line, underway, throughout the entire poem. The poem opens a while within language itself and invites man to dwell within that while.

Meditative thinking (*Andenken*) has never been popular in Western philosophy, which has mostly associated thinking with problem-solving, ordering, calculation, and relegated meditation to the sphere of religion. Yet Aristotle considered meditation on the first principles the most noble activity that could take place—not for the purpose of solving problems about them but rather to feel a kinship with them, to dwell in their proximity. This is a highly elite conception of thinking; it emphasizes the exalted solitude of the thinker and ignores the sufferings and bewilderments of the multitudes.

Heidegger regards the multitudes as stubborn, indifferent to the dignity of thought. Yet we cannot overlook the fact that thinkers always come forth out of this gray backdrop of mass humanity. The continuity of the fated mission of Being (*Seinsgeschick*) is therefore as much a function of mass humanity as of the single thinkers. I believe Heidegger has failed to stress this point sufficiently. He is usually highly sensitive to the originative contexts out of which things emerge and which sustain them. This is why earth is such an important category for him. The context out of which thinking emerges is the twofoldedness of Being: what-is/be-ing (*Seiendes/seiend*). But thinking cannot emerge without a thinker. Is it not mass humanity that sustains the little child who grows up to become a thinker? The analysis of *das Man* and conscience in *Sein und Zeit* provided a foundation for a philosophic pursuit of this idea; authentic *Da-sein* is hauled out of *das Man*. But the idea that *das Man* is therefore also a kind of ground or sustaining soil for thinkers to emerge out of is not really taken up.

If it is true that the thinker emerges out of *das Man*, then the thinker can only carry out his responsibility to the unfolding/folding by also assuming a responsibility to the multitudes. But to pinpoint the nature of this responsibility is a difficult problem. Should the thinker communicate with the masses and attempt to guide them? People at large are undisciplined; they tend to accept the results of thought without going through the pain of thinking. When this happens philosophy becomes for them ideology. Sometimes people hide the pain they feel when trying to think by ridiculing thought itself and labeling thinkers as "out of it." There appears to be a tragic conflict between the thinker and humanity at large. Yet the masses are also laid claim to by the Difference; the problem is that they do not openly acknowledge this. Perhaps the thinker should attempt to awaken them to an open response. But what sort of influence would this be?

Language speaks, not man. Human speech is a response to the silent interplay of language itself (SvGr 161). But then to whom does the thinker "speak"? If language, rather than man, speaks, then language turns out to be a monologue into which each man must singly acquiesce. The result appears to be a rather brutal picture of isolation, where men can turn toward Being and its empty withdrawal but they can't really turn toward one another. Drawn into the hazardous openness of the Region, each of us is surrounded by empty darkness. Yet it could be that Heidegger's position comes close to what is the case. If men are closer to the primordial silence and its breach than they are to each other, no wonder that philosophic thought wins few converts among humanity at large. Philosophic thought cannot persuade or it ceases to be authentic thought and instead degenerates into dogma and ideology. It is up to each man alone to discover his relation to the silent call; no one can do this for him. Markers along the way of thought can be highlighted, but each person himself must find the response to and the explication of these markers.

The ontological difference is for Heidegger much more than a theoretical problem. It is the basis for all ontic disparities and of all harmonies that grow out of these disparities. It applies to abstract, theoretical considerations such as the logic of presence, but also to practical, immediate problems of human understanding and human relationships. It cuts across the theoretical and the practical. Without paying heed to the Difference, even our dealings with ontic affairs are blind and shallow. To be sure, this blindness and shallowness is the mark of an everyday life into which we are all thrown. We cannot escape its facticity. But, by attending to the Difference and its silent call man can gain the requisite detachment from the ontic in order to exist ecstatically, to see through things while still accepting their factual presence. Man can leap. In this way the things themselves become transfigured and encountered in their higher presence, called forth and yet harboring an ensconcement. Only in this way can man come to terms with the alienation and distraction which cloud human existence and threaten the very status of man as an ecstatic, free being. To do so is to follow the other pole of our destiny. Although we are set adrift, we are first and foremost called forth. Only a few respond to the primordial call. They are the thinkers. They are the ones who pay heed to ". . . the two-fold of things-that-are and Being. That is what gives true thinking its substance and authentic character" (WhD 149).[4] And yet they too are underway, following in the wake of an event long with us: the simple fold has remained and continues to remain withdrawn from us precisely through ensconcing itself in the twofoldedness of open presence and things-present. Those who realize this are in the position of someone who suddenly finds out that someone he has known for a long time has appeared to him all the while in a disguise. What does it all mean? What are the consequences of this uncanny eclipse of Being which nevertheless releases and disguises itself into the colorful multiplicity of things-that-are? Can one be underway for an eternity? Is the "not" between Being and the things-that-are a "not-yet"? The gap could close between Being and the things-that-are, but would we be there? Or perhaps the withdrawal could become so decisive that all presence and all things-present are withdrawn? We can only wait. To wait is to dwell in the Between, the fold of the twofold. It is in waiting that we are.

Bibliographic Key to Textual References

Symbols used for text references to works by Martin Heidegger

EiM	Einführung in die Metaphysik. Tübingen: Niemeyer, 1953.
FD	Die Frage nach dem Ding: Zu Kants Lehre von den transzendentalen Grundsätzen. Tübingen: Niemeyer, 1962.
GD	"Grundsätze des Denkens." *Jahrbuch für Psychologie und Psychotherapie* (1958), pp. 33–41.
Gel	Gelassenheit. Pfullingen: Neske, 1959.
HEH	"Hölderlins Erde und Himmel." *Hölderlin-Jahrbuch* (1958–60), pp. 17–39.
HG	"Hegel and die Griechen." *Die Gegenwart der Griechen im neueren Denken. Festschrift for H. G. Gadamer.* Tübingen: Niemeyer, 1960, pp. 43–57.
Hö	Erläuterungen zu Hölderlins Dichtung. 2d ed. rev. Frankfurt: Klostermann, 1951.
Holz	Holzwege. Frankfurt: Klostermann, 1950.
Hum	Über den Humanismus. Frankfurt: Klostermann, 1949.
ID	Identität und Differenz. Pfullingen: Neske, 1957.
KM	Kant und das Problem der Metaphysik. 2d ed. Frankfurt: Klostermann, 1951.
KTS	Kants These über das Sein. Frankfurt: Klostermann, 1962.
1 Ni	Nietzsche, vol. 1. Pfullingen: Neske, 1961.
2 Ni	Nietzsche, vol. 2. Pfullingen: Neske, 1961.
PL	Platons Lehre von der Wahrheit. 2d ed. Bern: Francke, 1954.

SvGr	Der Satz vom Grund. Pfullingen: Neske, 1957.
SZ	Sein und Zeit. 8th ed. unaltered. Tübingen: Niemeyer, 1957.
TK	Die Technik und die Kehre. Pfullingen: Neske, 1962.
USp	Unterwegs zur Sprache. Pfullingen: Neske, 1959.
VA	Vorträge und Aufsätze. Pfullingen: Neske, 1954.
WBP	"Vom Wesen und Begriff der Physis: Aristoteles Physik B 1," *Il Pensiero*, vol. 3 (1958), pp. 129–56, 265–89. Reprinted in WgM, pp. 309–71.
WG	Vom Wesen des Grundes. 4th ed. Frankfurt: Klostermann, 1955.
WgM	Wegmarken. Frankfurt: Klostermann, 1967.
WhD	Was heisst Denken? Tübingen: Niemeyer, 1954.
WM	Was ist Metaphysik? 7th ed. Frankfurt: Klostermann, 1955
WP	Was ist das—die Philosophie? Pfullingen: Neske, 1956.
WW	Vom Wesen der Wahrheit. 3d ed. Frankfurt: Klostermann, 1954.
ZSf	Zur Seinsfrage. Frankfurt: Klostermann, 1956.

Primary Bibliography

In addition to those works found in the Bibliographic Key to Textual References, the following are germane to this study:

Martin Heidegger. *Der Feldweg*. Frankfurt: Klostermann, 1953

———. *Hebel—der Hausfreund*. Pfullingen: Neske, 1957

———. "Zu einem Vers von Morike." *Trivium* 9, no. 1 (1951): 1–16. (Correspondence between Heidegger and Emil Staiger.)

Secondary Bibliography

My list of secondary works is primarily confined to those most germane to the topics I have taken up in this book. However, I have also included selection monographs that I believe compensate for their generality by their extraordinary lucidity of freshness or that contain material directly relevant to the problem of the ontological difference.

Birault, Henri. "Existence et verite d'apres Heidegger." *Revue de Metaphysique et de Morale* 56 (1950): 35–87.

Borgmann, Albert. "Heidegger and Symbolic Logic." *Heidegger and the Quest for Truth*. Edited by Manfred S. Frings. Chicago: Quadrangle Books, 1968.

Dondyne, Albert. "La difference ontologique chez M. Heidegger." *Revue philosophique de Louvain* 56 (1958): 35–62, 251–93.

Harries, Karsten. "Heidegger: The Search for Meaning." *Existential Philosophers: Kierkegaard to Merleau-Ponty*. Edited by George A. Schrader. New York: McGraw-Hill, 1967.

Kockelmans, Joseph J. "Ontological Difference, Hermeneutics, and Language." Paper presented at Heidegger Symposium, The Pennsylvania State University, 1969.

Langan, Thomas. *The Meaning of Heidegger*. New York: Columbia University Press, 1959.

Liebrucks, Bruno. "Idee und ontologische Differenz." *Kantstudien*. vol. 48 (1956–57): H 12, 268–302.

Lingis, A.F. "On the Essence of Technique." *Heidegger and the Quest for Truth*. Edited by Manfred S. Frings. Chicago: Quadrangle Books, 1968.

Lohman, J. "M. Heideggers 'Ontologische Differenz' und die Sprache," *Lexis* 1 (1948): 49–106.

Löwith, Karl. *Heidegger: Denker in cürftiger Zeit*. Frankfurt: Fisher, 1953.

Macomber, W.B. *The Anatomy of Disillusion*. Evanston: Northwestern University Press, 1967.

Martin Heidegger zum siebzigsten Geburtstag. Festschrift. Pfullingen: Neske, 1959.

Marx, Werner. Heidegger und die Tradition. Stuttgart: Kohlhammer, 1961.

Palmer, Richard E. Hermeneutics. Evanston: Northwestern University Press, 1969.

Pöggeler, Otto. Der Denkweg Martin Heideggers. Pfullingen: Neske, 1963.

Pöggeler, Otto. "Sein als Ereignis." Zeitschrift für philosophische Forschung 13, no. 4 (October–December 1959): 597–632.

Richardson, William J. Heidegger: Through Phenomenology to Thought. The Hague: Nijhoff, 1963.

Versenyi, Laszlo. Heidegger, Being and Truth. New Haven: Yale University Press, 1965.

Voekel, Theodore S. "Heidegger and the Problem of Circularity." Ph. D. dissertation, Yale University, 1971.

Wiplinger, Fridolin. Wahrheit und Geschichtlichkeit. Freiburg: Alber, 1961.

Notes

Chapter I

1. William J. Richardson, *Heidegger: Through Phenomenology to Thought* (The Hague: Martinus Nijhoff, 1963), p. 15.

2. "Sinn ist das, worin sich Verständlichkeit von etwas hält" (*SZ* 151).

3. "Der Name nennt jenes, was wir meinen, wenn wir 'ist' sagen und: 'ist gewesen' und 'ist im Kommen.' Alles, was uns erreicht und wohin wir reichen, geht durch das gesprochene oder ungesprochene 'es ist' hindurch. Dass es sich so verhält—dem können wir nirgends und nie entgehen. Das 'ist' bleibt uns in allen seinen offenkundigen und verborgenen Abwandlungen bekannt."

4. Werner Marx, *Heidegger und die Tradition* (Stuttgart: Kohlhammer, 1961), p. 99.

5. Edmund Husserl, *Ideen zu einer reinen Phänomenologie und phänomenologischen Philosophie* (The Hague: Nijhoff, 1950), pp. 13, 16.

6. Etienne Gilson, *Being and Some Philosophers* (Toronto: Pontifical Institute of Mediaeval Studies, 1952), p. 4.

7. Husserl, *Ideen zu einer reinen Phänomenologie*, p. 13.

8. Macquarrie-Robinson renders this: "that which shows itself in itself" which is fine as long as there is some explanation of the phrase "in-itself" (*Being and Time*, p. 51).

9. "Als Bedeutung des Ausdrucks 'Phänomen' ist daher festzuhalten: das *Sich-an-ihm-selbst-zeigende*, das *Offenbare*. Die ... 'Phänomene' sind dann die Gesamtheit dessen, was am Tage liegt oder ans Licht gebracht werden kann, was die Griechen zuweilen einfach mit *ta onta* (das Seiende) identifizierten."

10. "Die Möglichkeit besteht sogar, dass Seiendes sich als das ziegt, was es an ihm selbst *nicht* ist. In diesem Sichzeigen 'sieht' das Seiende 'so aus wie. . . .' Solches Sichzeigen nennen wir *Scheinen*."

11. "Wieviel Schein jedoch, soviel 'Sein.'"

12. "Offenbar solches, was sich zunächst und zumeist gerade *nicht* zeigt, was gegenüber dem, was sich zunächst und zumeist zeigt, *verborgen* ist, aber zugleich etwas ist, was wesenhaft zu dem, was sich zunächst und zumeist zeigt, gehört, so zwar, dass es seinen Sinn und Grund ausmacht."

13. "Dass wir je schon in einem Seinsverständnis leben und der Sinn von Sein zugleich im Dunkel gehüllt ist, beweist die grundsätzliche Notwendigkeit, die Frage nach dem Sinn von 'Sein' zu wiederholen."

14. "Seinsverständnis meint hier [i.e., in *Sein und Zeit*, which Heidegger is discussing in this particular paragraph] niemals, der Mensch besitze als Subjekt eine subjektive Vorstellung vom Sein und dieses, das Sein sei eine blosse Vorstellung.... Seinsverständnis besagt, dass der Mensch seinem Wesen nach im Offenen des Entwurfes des Seins steht und dieses so gemeinte Verstehen aussteht."

15. "Sein lichtet sich dem Menschen im ekstatischen Entwurf. Doch dieser Entwurf schafft nicht das Sein.... Das Werfende im Entwerfen ist nicht der Mensch, sondern das Sein selbst, das den Menschen in die Ek-sistenz des Da-seins als sein Wesen schickt."

16. Wiplinger recognizes that *logos* and *phenomenon* belong in an inner unity, but he views both as grounded in *phainesthai*. But it is misleading to suggest that *logos* is in any way derivative of *phainesthai*; if anything, *logos* comes to be the deeper term, although it is not certain whether Heidegger, at the time he wrote *Sein und Zeit*, appreciated the status *logos* was to have for him. Fridolin Wiplinger, *Wahrheit und Geschichtlichkeit* (Freiburg: Alber, 1961), p. 141.

17. Wiplinger, *Wahrheit* (p. 141) quite correctly recognizes that there is a reflexive relation implicit in self-showing. But he proceeds to argue that self-showing shows itself *in itself*, i.e., within its own bounds. This follows only if it is assumed that *logos* is derivative of *phainesthai*. Wiplinger concludes, "Phenomenology is self-designation, self-interpretation of Being itself." ("Phänemenologie ist Selbstaufweis, Selbstauslegung des Seins selbst.") This interpretation would fit nicely for Hegel, but I think it involves difficulties in Heidegger's case. Phenomenology must not be understood as a *process* which Being undergoes in order to culminate in a self-interpretation. There is no guarantee that letting-be-seen results in a clarification of self-showing; in fact a concealment might result instead. If Being interprets itself, why hasn't it done so yet? Why does Being seem to need man if Being itself can interpret itself? Why man? Why not a simple, formal unity of the type Aristotle envisioned: thought thinking thought?

18. "... Sein eröffnet sich den Griechen als *physis* ... Die Wortstämme *phy-* und *pha-* nennen dasselbe. *Phyein*, das in sich ruhende Aufgehen ist *phainesthai*, Aufleuchten, Sichzeigen, Erscheinen."

19. "Im Wesen der Erscheinung liegt das Auf- und Abtreten, das Hin-und Her- in dem echt demonstrativen, zeigenden Sinne. Das Sein ist so in das mannigfaltige Seiende verstreut."

20. "Der *logos* der Phänomenologie des Daseins hat den Charakter des *hermeneuein*, durch das dem zum Dasein selbst gehörigen Seinsverständnis der eigentliche Sinn von Sein und die Grundstrukturen seines eigenen Seins kundgegeben werden. Phänomenologie des Daseins ist Hermeneutik in der ursprünglichen Bedeutung des Wortes, wonach es das Geschäft der Auslegung bezeichnet."

21. "Die Bedingungen der Möglichkeit der Erfahrung überhaupt sind zugleich Bedingungen der Möglichkeit der Gegenstände der Erfahrung."

22. "Im Wesen der Erscheinung liegt das Auf- und Abtreten, das Hin- und Her in dem echt demonstrativen, zeigenden Sinne. Das Sein ist so in das mannigfaltige Seiende verstreut."

23. "Beiläufig vermerkt, ist es kein Zufall, dass die griechische Sprache am deutlichsten und schärfsten spricht, wenn sie das von uns so genannte 'Seiende' im Plural des Neutrums nennt. Denn das Seiende ist ein jeweiliges und so ein vielfältiges; dagegen ist das Sein einzig, der absolute Singular in der unbedingten Singularität."

24. "Das Sein ist das Allgemeinste, was in jeglichem Seienden angetroffen wird und daher das Gemeinste, das jede Auszeichnung verloren oder noch nie besessen hat. Zugleich ist das Sein das Einzigste, dessen Einzigartigkeit von keinem Seienden je erreicht wird. Denn gegen jedes Seiende, das hervorragen möchte, steht doch noch immer seinesgleichen, d.h. immer Seiendes, wie verschiedenartig es auch bleiben mag. Das Sein aber hat nicht seinesgleichen."

Chapter II

1. "Weil das Sein, *physis*, im Erscheinen, im Darbieten von Aussehen und Ansichten besteht, steht es wesensmässig und somit notwentig und ständig in der Möglichkeit eines Aussehens, das jenes, was das Seiende in Wahrheit ist, d.h. in der Unverborgenheit, gerade verdeckt und verbirgt."

2. "Das umsichtig auf sein Um-zu Auseinandergelegte als solches, das ausdrücklich Verstandene, hat die Struktur des *Etwas als Etwas*."

3. "Sein west als *physis*. Das aufgehende Walten ist Erscheinen. Solches bringt zum Vorschein. Darin liegt schon: das Sein, Erscheinen, lässt aus der Verborgenheit heraustreten."

4. In contrast to the closure implied by doctrines of necessary truths the as-structure points up the openness of things and the fact that they *can be otherwise*. See W. B. Macomber, *The Anatomy of Disillusion* (Evanston: Northwestern University Press, 1967) pp. 93–103.

5. There is of course much more to be said. The duality of revealing/being-revealed is rooted in the more basic duality of revealing/concealing. A. Dondeyne apparently interprets the ontological difference as concerned with the former duality, revealing/being-revealed. Accordingly he interprets Heidegger as following in the mainstream of Kantian Transcendental Philosophy but deepening and widening it by attempting an ontology. Unfortunately Dondeyne does very little with the latter distinction which is really the basis for the former. The ontological difference surely involves both dualities. See A. Dondeyne, "La différence ontologique chez M. Heidegger," *Revue Philosophique de Louvain* 56 (1958), pp. 35–62, 251–93.

6. "Die Verborgenheit versagt der Aletheia das Entbergen ... bewahrt ihr das Eigenste als Eigentum.... Die Verborgenheit des Seienden im Ganzen ... ist älter als jede Offenbarkeit von diesem und jenem Seienden."

7. "Die Umgetriebenheit des Menschen weg vom Geheimnis hin zum Gangbaren, fort von einem Gängigen, fort zum nächsten und vorbei am Geheimnis, ist das Irren."

8. "Im Zugleich der Entbergung und Verbergung waltet die Irre."

9. Richardson (*Through Phenomenology*, p. 236) and Birault (in *Revue de Metaphysique et de Morale* 56 [1950], p. 71) connect being in error with what SZ had characterized as fallenness. I would say that fallenness is a symptom of error; fallenness emphasizes a conditioning-structure of human existence, whereas error comes to be regarded by Heidegger as dictated by the destiny of Being, revealing/concealing, itself. Cf. "Dergestalt beirrt das Sein, es lichtend, das Seiende mit der Irre" (*Holz* 310).

10. "In dem Schuhzeug schwingt der verschwiegene Zuruf der Erde, ihr stilles Verschenken des reifenden Korns und ihr unerklärtes Sichversagen in der öden Brache des winterlichen Feldes ... Dienlichkeit. Aber diese selbst ruht in der Fülle eines wesentlichen Seins des Zeuges ... die Verlässlichkeit. Kraft ihrer ist die Bäuerin durch dieses Zeug eingelassen in den schweigenden Zuruf der Erde, kraft der Verlässlichkeit des Zeuges ist sie ihrer Welt gewiss. Welt und Erde sind ihr und denen, die mit ihr in ihrer Weise sind, nur so da: im Zeug."

11. "... das, wohin das Aufgehen alles Aufgehende und zwar als ein solches zurückbirgt. Im Aufgehenden west die Erde als das Bergende."

12. "Zum Offenen gehört eine Welt und die Erde."

13. "Welt ist das immer Ungegenständliche, dem wir unterstehen, solange die Bahnen von Geburt und Tod, Segen und Fluch uns in das Sein entrückt halten. Wo die wesenhaften Entscheidungen unserer Geschichte fallen, da weltet die Welt. Der Stein ist weltlos. . . . Dagegen hat die Bäuerin eine Welt, weil sie sich im Offenen des Seienden aufhält. . . . Indem eine Welt sich öffnet, bekommen alle Dinge ihre Weile und Eile, ihre Ferne und Nähe, ihre Weite und Enge."

14. "Aber die Welt ist nicht einfach das Offene, was der Lichtung . . . entspricht. Vielmehr ist die Welt die Lichtung der Bahnen der wesentlichen Weisungen, in die sich alles Entscheiden fügt."

15. "Wahrheit west nur als der Streit zwischen Lichtung und Verbergung in der Gegenwendigkeit von Welt und Erde."

16. ". . . im Glanz west der Gott an. Im Abglanz dieses Glanzes glänzt, d.h. lichtet sich jenes, was wir die Welt nannten."

17. "Geschichte ist selten. Geschichte ist nur dann, wenn je das Wesen der Wahrheit anfänglich entschieden wird."

18. "Die Lichtung, in die das Seiende hereinsteht, ist in sich zugleich Verbergung . . . die offene Stelle inmitten des Seienden, die Lichtung, ist niemals eine starre Bühne mit ständig aufgezogenen Vorhang, auf der sich das Spiel des Seienden abspielt. Vielmehr geschieht die Lichtung nur als Verbergen."

19. "Der echte Anfang enthält schon verborgen das Ende."

Chapter III

1. "Wenn anders nun das Auszeichnende des Daseins darin liegt, dass es Seinverstehend zu Seiendem sich verhält, dann muss *das Unterscheidenkönnen*, in dem die ontologische Differenz faktisch wird, die Wurzel seiner eigenen Möglichkeit im Grunde des Wesens des Daseins geschlagen haben. Diesen Grund der ontologischen Differenz nennen wir vorgreifend die Transzendenz des Daseins."

2. "Das Verhältnis zum Seienden und der Bezug zum Sein/Die ontologische Differenz."

3. "Wir stehen in der Unterscheidung von Seiendem und Sein" (Original text italicized).

4. ". . . die Zwiefalt selber erst die Klarheit, d.h. die Lichtung entfaltet, innerhalb deren Anwesendes als solches und Anwesen für den Menschen unterscheidbar werden . . . für den Menschen, der seinem Wesen nach im Bezug, d.h. im Brauch der Zwiefalt steht."

5. Karsten Harries, "The Search for Meaning," *Existential Philosophers: Kierkegaard to Merleau-Ponty*, ed. George Alfred Schrader, Jr. (New York: McGraw-Hill, 1967), p. 171.

6. This is, of course, ambiguous; for on the one hand, the things-that-are demand our care and attention; but on the other hand, as *Sein und Zeit* proceeds to argue, human existence is the call of care itself. Thus that which lays claim to our essence can wear many clothes.

7. "Seinlassen . . . bedeutet, sich einlassen auf das Offene und dessen Offenheit, in die jegliches Seiende hereinsteht, das jene gleichsam mit sich bringt."

8. Wiplinger gives a fairly extensive analysis of the ontological significance of death within the Heideggerian thought framework. He connects the ability to die with the historical overcoming of subjectivity which makes possible an overcoming

of the metaphysical perspective (the forgetting of Being). See *Wahreit und Geschichtlichkeit* (Freiburg: Alber, 1961), pp. 240–47; 321–23.

9. "In seinem Tod muss sich das Dasein schlechtin 'zurücknehemen.' "

10. "Wenn wir zum Brunnen, wenn wir durch den Wald gehen, gehen wir schon immer durch das Wort 'Brunnen,' durch das Wort 'Wald' hindurch, auch wenn wir diese Worte nicht aussprechen und nicht an Sprachliches denken."

11. Versenyi, *Heidegger, Being and Truth* (New Haven: Yale University Press, 1965), p. 89, points out the similarity of this line of thinking to Kierkegaard's discussion of dispair in *The Sickness Into Death*. Dispair is the condition of a self which is opaque to its own finitude and refuses to admit that something else constitutes it.

12. The essay by Karsten Harries in *Existential Philosophers: Kierkegaard to Merleau-Ponty*, ed. George A. Schrader (New York: McGraw-Hill, 1967), is by far the best and most interesting "negative" interpretation of the tensions in human existence that I have read. Whereas in my own interpretation I concentrate on the resolution of tensions, he presents them as tensions with a dynamic lucidity. He reminds us that a decisive overcoming of the ontic standpoint would result in a being which is no longer in the world. This would mean madness or death. Hence the dire sacrifices of poets such as Hölderlin, who became hopelessly insane, or Trakl, who committed suicide (p. 202). Mr. Harries is not making a value judgment here; rather he points out that the "resolution" of the later Heidegger seems to involve a loss of the familiar everyday world and "man's daily struggle for life and love" (p. 200).

13. ". . . dichterisch, wohnet der Mensch auf dieser Erde."

14. "Wir können die technischen Gegenstände im Gebrauch so nehmen, wie sie genommen werden müssen. Aber wir können diese Gegenstände zugleich auf sich beruhen lassen als etwas, was uns nicht im Innersten und Eigentlichen angeht. Wir können 'ja' sagen zur umgänglichen Benutzung der technischen Gegenstände, und wir können zugleich 'nein' sagen, insofern wir ihnen verwehren, dass sie uns ausschliesslich beanspruchen und so unser Wesen verbiegen, verwirren und zuletzt veröden."

15. ". . . das 'es,' was hier 'gibt,' ist das Sein seibst. Das 'gibt' nennt jedoch das Gebende, seine Wahreit gewährende Wesen des Seins. Das Sichgeben ins Offene mit diesem selbst ist das Sein selber."

16. Heidegger includes Hegel among these thinkers and thus locates him in the Modern tradition. For a criticism of Heidegger's interpretation of Hegel, see Bruno Liebrucks, "Idee und Ontologische Differenz," *Kantstudien* 48, no. 12 (1956–57) pp. 268–302.

17. "Das Gedächtnis besagt ursprünglich soviel wie An-dacht: das unablässige, gesammelte Bleiben bei . . . und zwar nicht etwa nur beim Vergangenen, sondern in gleicher Weise beim Gegenwärtigen und dem, was kommen kann. Das Vereigenen das Gegenwärtige, das Kommende erscheinen in der Einheit eines je eigenen An-wesens."

Chapter IV

1. "Das Sein entzieht sich, indem es sich in das Seiende entbirgt. Dergestalt beirrt das Sein, es lichtend, das Seiende mit der Irre. . . . Ohne die Irre wäre kein Verhältnis von Geschick zu Geschick, wäre nicht Geschichte."

2. "Wir suchen das Griechische weder um der Griechen willen, noch wegen einer Verbesserung der Wissenschaft; nicht einmal nur der deutlicheren Zwiesprache halber, sondern einzig im Hinblick auf das, was in einer solchen Zwiesprache zur Sprache gebracht werden möchte, falls es von sich aus zur Sprache kommt. Das ist jenes Selbe, das die Griechen und uns in verschiedener Weise geschicklich angeht. . . . griechisch ist die Frühe des Geschickes, als welches das Sein selbst sich im Seienden lichtet und ein Wesen des Menschen in seinen Anspruch nimmt.

3. ". . . nach der Notwendigkeit; denn sie zahlen einander Strafe und Busse für ihre "Ungerechtigkeit."

4. ". . . entlang dem Brauch; gehören nämlich lassen sie Fug somit auch Ruch eines dem anderen (im Verwinden) des Un-fugs."

5. Richardson, in *Through Phenomenology to Thought*, translates this term as *mittence* (pp. 20–21, 493) to emphasize the way in which Being "sends itself" to where we are to be, thus constituting our situation as there, participating in a particular historical epoch. I do not follow Richardson because *mittence* has a theological ring which is not strictly implied by *Geschick*. Yet Richardson's translation has definite merit in that it links *Geschick* to disclosure or message and hence also to the primordial injunction (*Geheiss*).

6. "Dasselbe ist Denken und der Gedanke, dass Ist ist; denn nicht ohne das Seiende, in dem es als Ausgesprochenes ist, kannst du das Denken finden. Es ist ja nicht oder wird nichts anderes sein ausserhalb des Seienden, da es ja die *Moira* daran gebunden hat, ein Ganzes und unbeweglich zu sein."

7. "Es brauchet das Vorliegenlassen so (das) In-die-Achtnehmen auch: Seiendes seiend."

8. "Das Denken west der ungesagt bleibenden Zwiefalt wegen an. Das Anwesen des Denkens ist unterwegs zur Zwiefalt von Sein und Seiendem."

9. ". . . ihre Zwiefalt aus dem Verbergen ihrer Einfalt birgt das Geheiss."

10. "*Phasma* ist das Erscheinen der Sterne, des Mondes, ihr zum-Vorscheinkommen, ihr Sichverbergen. *Phasis* nennt die Phasen. Die wechselnden Weisen seines Scheinens sind die Mondphasen. *Phasis* ist die Sage; sagen heisst: zum Vorschein bringen."

11. "Die Entfaltung der Zwiefalt waltet als die *Phasis*, das Sagen als zum-Vorschein-bringen."

12. "Doch die Entbergung gewährt die Lichtung des Anwesens, indem sie zugleich, wenn Anwesendes erscheinen soll, ein vor-liegen-Lassen und Vernehmen braucht und also brauchend das Denken in die Zugehörigkeit zur Zwiefalt einbehält. Darum gibt es auf keine Weise ein irgendwo und irgendwie Anwesendes ausserhalb der Zwiefalt."

13. "[Die Zuteilung] . . . ist die in sich gesammelte und also entfaltende Schickung des Anwesens als Anwesen von Anwesendem. *Moira* ist das Geschick des 'Seins' im Sinne des *eon*."

14. "Im Geschick der Zwiefalt gelangen jedoch nur das Anwesen ins Scheinen und das Anwesende zum Erscheinen. Das Geschick behält die Zwiefalt als solche und vollends ihre Entfaltung im Verborgenen."

15. "Es ist ja nichts oder wird nichts anderes sein ausserhalb des Seienden, da es ja die *Moira* daran gebunden hat, ein Ganzes und unbeweglich zu sein. Darum wird alles blosser Name sein, was die Sterblichen so festgesetzt haben, es sei wahr:

'Werden' sowohl als 'Vergehen,' 'Sein' sowohl als 'Nicht-sein' und 'Verändern des Ortes' und 'Wechseln der leuchtenden Farbe.'"

16. "Dann waltet in der Entbergung ihr Sichverbergen? Ein kühner Gedanke. Heraklit hat ihn gedacht. Parmenides hat dies Gedachte ungedacht erfahren.

17. "Sie gehören wesenhaft zusammen auf Grund ihres Bezugs zum *Unterschied von Sein und Seiendem* (ontologische Differenz). Das dergestalt notwendig ontisch-ontologisch gegabelte Wesen von Wahrheit überhaupt ist nur möglich in eins mit dem Aufbrechen dieses Unterschiedes."

18. "Wie auch immer das Seiende ausgelegt werden mag, ... jedesmal erscheint das Seiende als Seiendes im Licht des Seins. Überall hat sich, wenn die Metaphysik das Seiende vorstellt, Sein gelichtet. Sein ist in einer Unverborgenheit (*Aletheia*) angekommen."

19. "Sein muss sogar von sich her und schon zuvor scheinen, damit jeweilig Seiendes erscheinen kann."

20. "Wie kann einer sich bergen vor dem, was nimmer untergeht?"

21. "Das griechisch gedachte Untergehen geschieht als Eingehen in die Verbergung."

22. "Das Aufgehen (aus dem Sichverbergen) dem Sichverbergen schenkt's die Gunst."

23. "Welt ist währendes Feuer, währendes Aufgehen nach dem vollen Sinne der *Physis*."

24. "... das lichtende Walten, das Weisen, das Masse gibt und entzieht."

25. "... das sinnend-versammelnde Vorbringer ins Freie, ist Gewähren von Anwesen."

26. The explanation of revealing/concealing given in *Vom Wesen der Wahrheit* is somewhat misleading to the extent that it overemphasizes the ontic realm of the entities and suggests that "something" is concealed, namely the totality of things-that-are, thereby implying that revealing and concealing might be separate occurrences. The most fundamental, incipient self-sameness of pure revealing/concealing per se is reflected not in the fact that "something" is concealed but rather in the austerity and forbidding emptiness of the open realm of presence.

27. "Ist diese Auszeichnung gar von einer Art, dass der Spruch solches erfrägt, was unausgesprochen auch jenes Anwesende zu sich einholt und bei sich einbehält, das zwar gebietsmässig nicht mehr unter die Menschen und die Götter zu rechnen, aber gleichwohl in einem anderen Sinne göttlich und menschlich ist, Gewächs und Getier, Gebirg und Meer und Gestirn?"

28. "... in das Ereignis der Lichtung vereignet, darum nie verborgen sondern ent-borgen ... (d.h.) der bergenden, sie haltenden und verhaltenden Lichtung zugetraut."

29. "... kaum wesend und nicht zurückgehend in den Anfang, sondern fortgehend in die blosse Unverborgenheit."

30. "Das alltägliche Meinen sucht das Wahre im Vielerlei des immer Neuen, das vor ihm ausgestreut wird. Es sieht nicht den stillen Glanz (das Gold) des Geheimnisses, das im Einfachen der Lichtung immerwährend scheint."

31. Studies such as Bruno Liebrucks' "Idee und Ontologische Differenz," *Kantstudien* 48, no. 12 (1956–57): 208–302, attempt to show that the Difference was

not completely forgotten. This particular study is provocative in that it looks at the ontological difference from the standpoint of Hegelian philosophy. However there are two difficulties with the essay: (1) The author does not grasp the role of concealment in the ontological difference. (2) As a result, he does not see that Hegel's awareness of the Difference must have been very shadowy, for he failed to grasp its full implications or his philosophy would have turned out very differently. I would argue that Hegel grasped the theme of transcendence but he did not grasp the inner meaning of *aletheia* as revealing/concealing.

32. I do not deal in any detail with Heidegger's interpretations of philosophers located within the metaphysical tradition, since this has been treated by such writers as Richardson, Dondeyne, Versenyi, Macomber and Liebrucks. Dondeyne focuses on the role the ontological difference plays as a principle of interpretation. However he claims that Heidegger himself operates within the grand tradition of *philosophia perennis* insofar as Heidegger acknowledges that a number of philosophers have concerned themselves with the riddle of there being things-that-are (for example, Aristotle, Leibniz, Kant). Unfortunately this view of Heidegger tends to suppress the eschatological character of his thought, that is, the problem of "overcoming" the tradition which Nietzsche brought to its end. Cf. Dondeyne, Albert, "La difference ontologique chez M. Heidegger," *Revue Philosophique de Louvain* 56 (1958): 35–62, 251–93.

33. "Sein muss sogar von sich her und schon zuvor scheinen, damit jeweiling Seiendes erscheinen kann."

34. "Sein west zwar als *physis*, als Sichentbergen, von sich her Offenkundiges, aber dazu gehört ein Sichverbergen. Fiele die Verbergung aus und weg, wie sollte dann noch Entbergung geschehen?"

35. "Die *Aletheia*—kaum wesend und nicht zurückgehend in den Anfang, sondern fortgehend in die blosse Unverborgenheit—kommt unter das Joch der Idee."

36. Many modern interpreters of Plato undoubtedly feel that the interpretation given here is a distorted view of Platonic philosophy. There is much evidence that the static interpretation of Platonic ideas has been overdone, especially in light of Plato's later dialogues. Plato may have had something in mind more of the character of hidden, regulative essences which shine forth when men inquire together in dialectic rather than manifesting themselves with final, dogmatic whatness. Nevertheless the historically dominating interpretation of Platonic ideas has been the one stressing static whatness, and since it is this historically dominating interpretation which has shaped the metaphysical tradition Heidegger is in this sense justified.

37. Pöggeler, Otto, *Der Denkweg Martin Heideggers* (Pfullingen, 1963), pp. 145–163.

38. "... sowohl das Sein als auch das Seiende je auf ihre Weise aus der *Differenz* her erscheinen."

39. "... Sein des Seienden heisst: Sein, welches das Seiende ist. Das 'ist' spricht hier transitiv, übergehend. Sein west hier in der Weise eines Uberganges zum Seienden. Sein geht jedoch nicht, seinen Ort verlassend, zum Seienden hinüber, so als könnte Seiendes, zuvor ohne das Sein, von diesem erst angegangen werden. Sein geht über (das) hin, kommt entbergend über (das), was durch solche Uberkommnis erst als von sich her Unverborgenes ankommt. Ankunft heisst: sich bergen in Unverborgenheit: also geborgen anwähren: Seiendes sein."

40. "Sein geht über (das) hin, kommt entbergend über (das), was durch solche Uberkommnis erst als von sich her Unverborgenes ankommt."

41. "Sein im Sinne der entbergenden Überkommnis und Seiendes als solches im Sinne der sich bergenden Ankunft wesen als die so Unterschiedenen aus dem Selben, dem Unter-Schied."

42. "Dieser [the Difference as simple fold] vergibt erst und hält auseinander das Zwischen, worin Überkommnis und Ankunft zueinander gehalten, auseinanderzueinander getragen sind."

43. "Er ist al äusserste Möglichkeit des sterblichen Daseins nicht Ende des Möglichen, sondern das höchste Ge-birg (das versammelnde Bergen) des Geheimmnisses der rufenden Entbergung."

44. "Unzählig ist das Vergehende und Vergangene, selten das Gewesene, seltener noch sein Gewähren."

45. Richardson in *Heidegger: Through Phenomenology to Thought* translates *Ereignis* as *e-vent* (p. 614) but calls our attention to the connection between *Ereignis* and *sich eignen*. This has merit insofar as it emphasizes the historical character of this term and the affinity between e-vent and *Austrag*. But e-vent does not capture the unity between that-which-takes-place-to-be-viewed and the viewing which lets it take place. Mr. Voelkel uses "self-designating appropriation," and the influence of his translation on mine is obvious. He offers one of the most thorough and most lucid treatments of this notion available in English in his dissertation *Heidegger and the Problem of Circularity* (New Haven: Yale University Press, 1971).

46. "Sein schickt sich dem Menschen zu, indem es lichtend dem Seienden als solchem einen Zeit-Spiel-Raum einräumt. Sein west als solches Geschick, als Sichentbergen, das zugleich währt als Sichverbergen."

47. "Die Zeitlichkeit 'ist' überhaupt kein *Seiendes*. Sie ist nicht, sondern zeitigt sich."

48. "Nichts geschieht, *das Ereignis er-eignet*. Der Anfang nimmt—austragend die Lichtung—den Abschied an sich."

49. "... sich uns zuschickt, indem es sich entzieht."

Chapter V

1. "Sein 'ist' im Wesen: Grund. Darum kann Sein nie erst noch einen Grund haben, der es begründen sollte."

2. There is an ambiguity surrounding interpreters such as Versenyi or Langan who find the position of the later Heidegger "mystical" and "unsatisfying." Is Versenyi's "Back to Humanism" a heroic insolence towards the Difference—a deliberate half-hearing—or is it just another lapse into *Vergessenheit*?

3. Karsten Harries has reminded me that the historical sense of "back" must also be emphasized here. In *Sein und Zeit* Heidegger treated the past as that which we come back to, our facticity. By stepping back we do not enter the dead realm of what is gone by but rather we locate ourselves openly within the momentous self-spectacle which gathers together past and even future as the self-ensconcing Mystery. It is in this Region that the thinker can have a vibrant encounter with previous thinkers: for only from within the Region does the fateful character of their thought become intelligible to us as responses to the Mystery and its primordial injunction, forgotten though it be. As Macomber points out, there is a Freudian bent to Heidegger's thought. Only a standpoint which embraces the hidden source of the surface meanings can give rise to a real interpretation. For Freud it is the

Region of the unconscious, where forgotten things still make their presence felt. For Heidegger it is the Region of the twofold and its Mystery. (B. Macomber, *The Anatomy of Disillusion*, Northwestern, 1967, p. 112.)

4. "... Massgabe und Bezirk (sind) nicht zwei verschiedene oder gar getrennte Sachen, sondern eine und dieselbe. Die Massgabe ergibt und öffnet einen Bezirk worin die Massgabe zu Hause ist und das sein kann, was sie ist."

5. "Das Wohnen des Menschen beruht im aufschauenden Vermessen der Dimension, in die der Himmel so gut gehört wie die Erde."

6. "So erscheint der unbekannte Gott als der Unbekannte durch die Offenbarkeit des Himmels. Dieses Erscheinen ist das Mass, woran der Mensch sich misset."

7. "Was dem Gott fremd bleibt, die Anblicke des Himmels, dies ist dem Menschen das Vertraute. Und was ist dies? Alles, was am Himmel und somit unter dem Himmel, und somit auf der Erde glänzt und blüht, tönt und duftet, steigt und kommt, aber auch geht und fällt, aber auch klagt und schweigt, aber auch erbleicht und dunkelt. In dieses dem Menschen Vertraute, dem Gott aber Fremde, schicket sich der Unbekannte, um darin als der Unbekannte behütet zu bleiben."

8. "Das dichtende Sagen der Bilder versammelt Helle und Hall der Himmelserscheinungen in Eines mit dem Dunkel und dem Schweigen des Fremden."

9. "Der Himmel ist der wölbende Sonnengang, der gestaltwechselnde Mondlauf, der wandernde Glanz der Gestirne, die Zeiten des Jahres und ihre Wende, Licht und Dämmer des Tages, Dunkel und Helle der Nacht, das Wirtliche und Unwirtliche der Wetter, Wolkenzug und blauende Tiefe das Äthers."

10. "Der Glanz des Himmels ist Aufgang und Untergang der Dämmerung, die alles Verkündbare birgt. Dieser Himmel ist das Mass."

11. "Dieses Denken ist im Begriff, die Erde als Erde preiszugeben."

12. "... die dienend Tragende, die blühend Fruchtende, hingebreitet in Gestein und Gewässer, aufgehend zu Gewächs und Getier."

13. "Aber die wahre Zeit ist Ankunft des Gewesenen. Dieses ist nicht das Vergangene, sondern die Versammlung des Wesenden, die aller Ankunft voraufgeht, indem sie als solche Versammlung sich in ihr je Früheres zurückbirgt."

14. "... das Gewesene. Damit meinen wir die Versammlung dessen, was gerade nicht vergeht, sondern west, d.h. währt, indem es dem Andenken neue Einblicke gewährt."

15. Pöggeler points out that this part of Heidegger's thinking is no less historical than his attention to the metaphysical tradition and the early Greek thinkers. The oldest traditions of *Mythos* conceive of the world as this foursome. See Otto Pöggeler, *Der Denkweg Martin Heideggers* (Pfullingen: Neske, 1963), p. 248.

16. "Sobald menschliches Erkennen hier ein Erklären verlange, übersteigt es nicht das Wesen von Welt, sondern es fällt unter das Wesen von Welt herab. Das menschliche Erklärenwollen langt überhaupt nicht in das Einfache der Einfalt des Weltens hin."

17. "Als so Einige sind sie innig. Die Mitte der Zwei ist die Innigkeit."

18. "Dass Welt sich ereigne als Welt, dass dinge das Ding, dies ist die ferne Ankunft des Seins selbst."

19. This point is made in a different setting by Albert Bergman, "Heidegger and

Symbolic Logic" *Heidegger and the Quest for Truth*, ed. Manfred S. Frings (Chicago: Quadrangle Books, 1968), pp. 139–58. Cf. pp. 154–55.

20. To be more specific, the primordial injunction manifests itself as Need, and technology is the response to the call of Need. This point and the implied precarious relation between language and technology is discussed by A. F. Lingis, "On the Essence of Technique," *Heidegger and the Quest for Truth*, ed. Frings, pp. 126–37.

21. "Das Wesen der modernen Technik beruht im Ge-Stell. Dieses gehört in das Geschick der Entbergung."

22. "Welche Anwesenheit ist die höhere, die des Vorliegenden oder die des Gerufenen?"

23. ". . . nicht zurückgehend in den Anfang, sondern fortgehend in die blosse Unverborgenheit . . ."

Chapter VI

1. "Das erbringende Eignen, das die Sage als die Zeige in ihrem Zeigen regt, heisse das Ereignen. Es er-gibt das Freie der Lichtung, in die Anwesendes anwähren, aus der Abwesendes entgehen und im Entzug sein Währen behalten kann."

2. "Die im Ereignis beruhende Sage ist als das Zeigen die eigenste Weise des Ereignens. Das Ereignis ist sagend."

3. "Am Ende muss sich die philosophische Forschung einmal entschliessen zu fragen, welche Seinsart der Sprache überhaupt zukommt. Ist sie ein innerweltlich zuhandenes Zeug, oder hat sie die Seinsart des Daseins oder keines von Beiden?"

4. It should be noticed however that the later writings of Heidegger—for instance, *Holzwege*, *Was heisst Denken?*—do not fasten on an analysis of tool being to explicate the craftsman, but rather on his relationship to his materials and the element which is their source. A woodcraftsman, above all, must have a special relationship to wood and the figures which slumber within it (WhD 54).

5. "Die Sprache ist das Haus des Seins. In ihrer Behausung wohnt der Mensch."

6. "Das Hören konstituiert sogar die primäre und eigentliche Offenheit des Daseins für sein eigenstes Seinkönnen, als Hören der Stimme des Freundes, den jedes Dasein bei sich trägt. Das Dasein hört, weil es versteht. Als verstehendes In-der-Welt-sein mit den Anderen ist es dem Mitdasein und ihm selbst 'hörig' und in dieser Hörigkeit zugehörig."

7. W. B. Macomber points out that breakdowns themselves bring a "moment of truth," much like the encounter with Nothingness, which vanishes almost immediately as once more the practical problem of repairing the breakdown sets in. The blindness of positivism parallels the blindness of the everyday existence. Both regard the vanishing glimpse as a mere subjective impression, rather than as a disclosure. See Macomber, *The Anatomy of Disillusion* (Evanston: Northwestern University Press), pp. 44–51.

8. There is a connection between the silence of the call and the breakdown of an instrumental complex. The silence could be considered as the disclosure that comes when language "breaks down." Cf. Richard E. Palmer, *Hermeneutics* (Evanston: Northwestern University Press, 1969), pp. 132–35. I will discuss this more in detail in the next section.

9. ". . . jenes Darlegen, das Kunde bringt, insofern es auf eine Botschaft zu hören vermag."

10. "Das Geschick des Seins ist als Zuspruch und Anspruch der Spruch, aus dem alles menschliche Sprechen spricht. Spruch heisst lateinisch fatum. Aber das Fatum ist als der Spruch des Seins im Sinne des sich entziehenden Geschickes nichts Fatalistisches . . ."

11. "Wenn wir bei der Sprache anfragen, nämlich nach ihrem Wesen, dann muss uns doch die Sprache selber schon zugesprochen sein. Wollen wir dem Wesen, nämlich der Sprache, nachfragen, so muss uns auch, was Wesen heisst, schon zugesprochen sein."

12. Heidegger uses the term be-wëgen to mean the clearing of a way or path where, presumably, there was none before. This word is based on wëgen, which means to blaze a path in the Alemannian/Swabian dialect (USp 261).

13. The connection between the "moment of truth" and the breakdown of an instrumental complex is still with us. But this flickering moment remains just that unless the everyday orientation is overcome.

14. "Allein, wann immer und wie immer wir eine Sprache sprechen, die Sprache selber kommt dabei gerade nie zum Wort."

15. "Wo aber kommt die Sprache selber als Sprache zum Wort? Seltsamerweise dort, wo wir für etwas, was uns angeht, uns an sich reisst, bedrängt oder befeuert, das rechte Wort nicht finden. Wir lassen dann, was wir meinen, im Ungesprochenen und machen dabei, ohne es recht zu bedenken, Augenblicke durch, in denen uns die Sprache selber mit ihrem Wesen fernher und flüchtig gestreift hat."

16. ". . . sagan heisst zeigen: erscheinen lassen, lichtend-verbergend frei-geben als dar-reichen dessen, was wir Welt nennen. Das lichtend-verhüllende, schleiernde Reichen von Welt ist das Wesende im Sagen."

17. "Das Wesende der Nähe ist nicht der Abstand, sondern die Bewëgung des Gegen-einander-über der Gegenden des Weltgeviertes."

18. "Wir nennen das lautlos rufende Versammeln, als welches die Sage das Welt-Verhältnis be-wëgt, das Geläut der Stille. Es ist: die Sprache des Wesens."

19. "Das Spiel ist ohne 'Warum.' Es spielt, dieweil es spielt. Es bleibt nur Spiel: das Höchste und Tiefste."

20. ". . . das Wort sei Wink und nicht Zeichen im Sinne der blossen Bezeichnung."

21. "So lernt ich traurig den Verzicht:
Kein Ding sei wo das Wort gebricht."

22. "Dies sind Worte, durch die das schon Seiende gemacht wird, dass es fortan glänzt und blüht und so überall im Lande als das Schöne herrscht."

23. "Welche Anwesenheit ist die höhere, die des Vorliegenden oder die des Gerufenen?"

24. It is this feature of language which is most drastically neglected by Lohmann and Kockelmans, thus making their treatment of "Language and Difference" one-sided. It is also significant that they do not emphasize the aspect of withdrawal in their explications of the ontological difference itself. Cf J. Lohman, "M. Heideggers Ontologische Differenz und Die Sprache," Lexis 1 (1948): 49–106, & J. J. Kockelmans "Ontological Difference, Hermeneutics, and Language" (Paper

presented at the Heidegger Symposium, The Pennsylvania State University, University Park, Pa., September 1969).

25. "Wort und Ding sind verschieden, wenn nicht geschieden."
26. "Das Wort: das Gebende. Was denn? Nach der dichterischen Erfahrung und nach ältester Überlieferung des Denkens gibt das Wort: das Sein."
27. "Die Bedingung ist der seiende Grund für etwas Seiendes. . . . Aber das Wort be-gründet das Ding nicht. Das Wort lässt das Ding als Ding anwesen."
28. "Die Sprache west als der sich ereignende Unter-Schied für Welt und Dinge."
29. "Das verlautende Wort kehrt ins Lautlose zurück, dorthin, von woher es gewährt wird:: In das Geläut der Stille . . ."
30. Theodore S. Voelkel, "Heidegger and the Problem of Circularity" (Ph.D. diss., Yale University, 1971), pp. 235–39, 246 ff.
31. "Eigentliche Dichtung ist niemals nur eine höhere Weise der Alltagssprache. Vielmehr ist umgekehrt das alltägliche Reden ein vergessenes und darum vernutztes Gedicht, aus dem kaum noch ein Rufen erklingt."
32. Versenyi, L., *Heidegger, Being and Truth* (New Haven: Yale University Press, 1965) p. 171.
33. This is what I find most disturbing about Versenyi's book. If Heidegger is right, on what grounds can we validly object to his thought? Versenyi attacks on normative grounds what is not really a normative argument. The intuitionistic character of Heidegger's last works is, I believe, exaggerated by Versenyi. These works Heidegger has carefully prepared for and led up to and we cannot extract them from Heidegger's entire *Denkweg*. Furthermore, as K. Harries remarks, "It takes strength and courage to wait" ("Heidegger, The Search for Meaning," p. 203).

Chapter VII

1. *Kein* Weg des Denkens, auch nicht der des metaphysischen, *geht* vom Menschenwesen aus und von da zum Sein über oder umgekehrt vom Sein aus und dann zum Menschen zurück. Vielmehr *geht* jeder Weg des Denkens immer schon *innerhalb* des ganzen Verhältnisses von Sein und Menschenwesen, sonst ist es kein Denken.
2. "Und so ist der Mensch, als existierende Transzendenz überschwingend in Möglichkeiten, ein Wesen der Ferne. Nur durch ursprüngliche Fernen, die er sich in seiner Transzendenz zu allem Seienden bildet, kommt in ihm die wahre Nähe zu den Dingen ins Steigen. Und nur das Hörenkönnen in die Ferne zeitigt dem Dasein mit dem es die Ichheit darangeben kann, um sich als eigentliches Selbst zu gewinnen."
3. "Die Ros ist ohn warum; sie blühet, weil sie blühet,
   Sie acht nicht ihrer selbst, fragt nicht, ob man sie siehet."
4. ". . . die Zwiefalt von Seiendem und Sein. Sie ist das, was eigentlich zu denken gibt."

Index

abidance, 149, 152
   as dwelling, 62 ff., 115
aletheia, 105 ff.
apophansis, 14, 28 ff., 134
arrival (Ankunft), 117 ff., 138
"as," 21, 29 f., 32, 87, 115
authentic/inauthentic, 61 ff., 183 f.

breach of stillness, 179 ff., 190, 203
brookage, 84 ff.

call (Ruf), 59, 178 ff. See also primordial injunction
closeness/remoteness, 73 ff., 80, 194 ff.
concealment, 16, 18, 33 ff., 54, 98, 106 ff., 192

death, 57 f.
decisive outbearance (Austrag), 118 ff., 144 ff.
Difference, 105, 113 ff., 146
dimension, 129 ff., 177 ff.
drift, 45 f., 83, 98, 112, 122, 189, 195
dwelling (Wohnen), 38, 62, 75, 132 f., 163 ff., 188

earth, 37 ff., 72, 136 f.
ecstase, 16, 49 ff., 117
ensconcement, 106, 117 ff., 133 f. See also concealment; revealing/concealing; unfolding/folding
Ereignis. See self-spectacle
error, 35, 45, 112
essence, 162, 166 ff.
everyday, 54 ff., 104 f., 120, 143, 184

fate/destiny (Geschick, Seinsgeschick), 7, 45, 94 ff., 98 ff., 103 ff., 121, 135, 168, 183. See also moira
favor (Gunst), 108. See also giving, gift
foursome (Geviert), 137, 140 ff.

gatherage, 92 ff., 101, 119, 139, 142
giving, gift, 41 ff., 87 ff., 142 f., 176

god, gods, 110, 135
ground, 126

happening (Geschehen), 40, 43 ff., 85. See also self-spectacle
   truth as happening, 40 ff.
heaven, 136
heeding, 78, 93 ff.
hermeneutic, 13 ff., 28 ff., 37, 133, 190
history, 43, 83, 112 f., 119 ff.
holy, 42, 62, 110, 133 f.

illuminative clearing (Lichtung), 40 f., 44, 48, 102 ff., 109, 118
   happens as concealing, 44
innerness, 144, 153
inter-cision (Unter-Schied), 144

juncture (Fug), 86 ff.

leap (Sprung), 68, 128
letting-be (Seinlassen), 51 ff., 64, 70, 74, 96
letting-be-seen, 14
letting-lie (Liegenlassen), 92 ff.
logos, 69, 91 ff., 101 ff., 166 ff.
   as letting-be-seen, 14

meaning, 28, 30, 36
   sense (Sinn) vs. referential meaning (Bedeutung), 6, 11
measure (Mass), 108 f., 130 ff.
moira, 103 ff.
Mystery (Geheimnis), 23, 35, 64, 68, 133, 199. See also error; concealment; holy

Nothingness, 125 ff.

ontology, 4 ff.
openness, 27, 32, 189

phainesthai, 15, 69, 166
phenomenon, 9, 14 ff.
physis, 15 f., 25, 38, 107 f., 114, 174, 200
poet, poetry, 133 ff., 151 ff.

presence, higher and lower sense of, 150 ff., 157, 190, 193 f.
presence/things-present, 87 ff., 101 ff., 106, 113 ff.
primordial injunction (anfänglicher Geheiss), 72, 74, 78, 104, 148, 156, 189, 195 ff.

reduction, 127, 190
Region (Gegend), 156 ff., 194
revealing/concealing, 31 ff., 44, 66, 105, 178
rift, 44

Saga, sagan, 170
Seinsverständnis, 12 ff., 83
self-spectacle, spectacle on-going (Ereignis), 120 ff., 128, 141, 157, 190
silence (Stille), 165, 174 ff., 202 f. See also withdrawal
 as stilling of world and thing, 179

spoken-to (Zuspruch), 169
supervention (Überkommnis), 116

technology, 68 ff., 146 ff.
thing, 142 ff., 174 ff.
thinking, 68 ff.
time, 33, 66 ff., 121
transcendence, 23, 32, 45, 47 ff., 61
Truth, 29, 40 ff.
twofold (Zwiefalt), 47, 99 ff., 115

underway, 71, 73, 82, 100, 198 ff.
unfolding, 101 ff., 170
unfolding/folding, 111, 144 f., 197

waiting, 198
while (Weile), 57, 86, 89, 200 f. See also dwelling
withdrawal (Entzug), 73, 112, 169 ff., 175 ff.
world, 39 ff., 50 f., 139 ff.